OPEN TAP DRINK POISON

LET'S FIX IT

By

TOM VOLINCHAK

Edited by Cathy Ann Simmons
My beloved wife, life saver, and partner forever. I love you, my dear!

Layout Design by Furkan Süperdoğan

ISBN: 978-1-64153-451-2

https://wateristhenewgold.com/

To Doctor Chester A. Amedia, MD – For charging me with bringing safe water to our dialysis patients. Your trust taught me the great responsibility that comes with treating water.

To Bill Watson – My first and greatest industry mentor. Your generous teaching set me off upon a wonderful thirty-year journey fixing water in ways that I could have never imagined.

Table Of Contents

1. Introduction

Safe water for everyone is a tenet of our existence and it frequently raises the debate of whether this precious resource should be a human right as opposed to being a commodity more readily available to those enjoying higher economic status. The thought that poor and impoverished people would have less access to safe water is repulsive to us, yet globally this is exactly the case with some nations suffering disease and economic despair because of the lack of fresh water.

Sadly, these kinds of problems are not isolated to poor third world countries. Right here in the US we have towns where aquifers, rivers and lakes have expired, where deadly toxins have poisoned local drinking supplies, and even situations where residents suffer illness and a loss of property for lack of the financial clout to fight for their water rights in court.

Let us step back for a moment and ponder just how important safe water is.

We drink water, we cook with it, we use it to grow crops, we make medicines with it, we manufacture and invent technology with it, we wash with it, heal with it, and we frolic in it. In short, water not only sustains our lives, but in many ways it makes life worth living.

Science also tells us of the importance of water. Life literally cannot exist without water. Not animal life, not plant life, not your life and not mine. Water is so necessary to life that NASA deems it as the mandatory ingredient to determine if another planet can support life. Water is so important that there are volumes of scientific journals explaining that when life first arose on earth, it arose from the sea; it arose from our magnificent water.

It is a grim and chilling reality then that one day soon we may not have safe drinking water. It does not matter where we live, it does not matter how plentiful our local water supply is, and it does not matter how smart our local water officials are. Our water is under attack and every day we face the real possibility of opening our faucet and having no water or having poisoned water. The threats to safe drinking water are real, abundant, formidable, and dangerous.

Drought, climate change, operator error, industrial and agricultural pollution, environmental spills, potential terrorism, aging infrastructure, pollution caused by our own individual daily activities, and more constantly hammer away at the safety of our water. We even have government agencies, communities and neighbors fighting legal battles over who our water belongs to.

Alarmingly, a simple Internet search of water contamination, shortage, and legal battles returns enough results to set off fear for our water. Here are just a few examples:

- In 2018, brought on by severe drought, Cape Town, South Africa, was on the very brink of experiencing Day Zero, the point where the city had no more water.[1]

1 https://www.mckinsey.com/business-functions/sustainability/our-insights/water-a-human-and-business-priority.

- In April 2014, the negligence of local officials in Flint, Michigan, caused its entire water supply to be contaminated with life-threatening lead contamination.
- On August 5, 2015, the EPA admitted to accidentally spilling 1 million gallons of water containing heavy metals into a tributary of the Colorado River.[2]
- Recent studies have uncovered extensive contamination of our water supplies with cancer-causing substances called PFAS. Labeled "forever chemicals," these compounds are used in the modern manufacturing of products such as nonstick cookware. Most of our aging treatment plants do not have the technology to remove this growing problem.[3]
- As many as 45 million people per year drink local water that is in violation of the Safe Drinking Water Act.[4]
- Two billion people, roughly 25% of the population, drink from water supplies that contain feces.[5]
- The American Society of Civil Engineers recently rated the overall condition of the US water infrastructure, including pipes, pumps, plants and more, at a D+ grade.[6]

We are far past the point where we can relax and feel comfortable knowing that safe drinking water is guaranteed to us. We are now at

2 https://www.cnn.com/2015/08/09/us/colorado-epa-mine-river-spill/index.html.

3 https://www.cnbc.com/2020/01/22/us-drinking-water-widely-contaminated-with-forever-chemicals-report.html.

4 https://www.sciencemag.org/news/2018/02/millions-americans-drink-potentially-unsafe-tap-water-how-does-your-county-stack#:~:text=In%20any%20given%20year%20from,rural%2C%20low%2Dincome%20areas.

5 https://www.who.int/news-room/fact-sheets/detail/drinking-water.

6 https://www.waterworld.com/drinking-water/article/16223567/us-infrastructure-gets-near-failing-grade-in-new-asce-report.

the point where we must assume that our water has an increasingly high probability of being contaminated, and we must from this point forward assume personal responsibility for our own safe drinking water.

Fortunately, there is an abundance of good news for us.

Technology companies have provided an endless selection of personal water purification products that can handle most of the contamination problems we are likely to face. Admittedly the sheer volume and claims of these products are mind-boggling.

What we have not had, at least up until now, is a thorough source of information and education to help us make proper water decisions.

I am proud to say that I have written *Open Tap – Drink Poison* for this very purpose.

The information within this book is varied, diverse and expansive. It applies to virtually every water challenge that you are likely to encounter on your own personal water journey. Some of the information is technical, some of it is common sense, some of it is secret-revealing and some of it I hope strikes you as humorous.

Nonetheless, all of it is designed to give you a starting place, a good water toolbox so that you can help those you love and care about to always have safe drinking water.

In this book we will talk about the basic nature of water, its chemistry, and physical properties. We will learn how to read a water analysis. We will learn the basic types of water contaminants and which filters/technologies are the best for us. We will study bottled water, specialty waters, home filtration products, water product health claims, and we will talk about boiling, distillation and what exactly constitutes pure water.

We will discuss hot topics like lead, arsenic, selenium, microplastics, oxy-anions, bacteria, viruses, fuel oils, PFAS, and chemicals with impossible-to-pronounce names, which sadly are becoming household words.

We will also talk about our aging pipe infrastructure including many failing treatment plants that were built as long ago as WWII but are still in operation. We will talk about climate change, drought, and even cover the many career and educational opportunities that exist in the water industry. We will talk about the lack of trained water professionals going into the future. In a blurb I call "The Great Water Grabber," we will even talk about how our greed is fast becoming a threat to safe water for everyone.

Finally, we will discuss America's dire need to develop a real national water policy.

So, without further introduction, I present to you my book *Open Tap – Drink Poison.*

I hope that this book touches you and helps you to gain a new understanding of water. I hope it enables you to understand what your local water supplier is doing, which water products are best for you, and which ones are not. I hope it gives you clear, safe direction should you ever face water contamination issues yourself. I hope that as you read it, you will, like I did all during my career, find new wonders about water, and perhaps even say to yourself, "I never knew that."

If my book brings you to understand, appreciate and love water even just a little bit more, then I will rest well knowing that there was a noble reason for the career I pursued, and for writing this book.

2.
The Nature of Water

From the very beginning of history continuing through today, humankind has spoken passionately about water and its mystical, magical properties. Ramses believed the waters of the Nile had healing powers, Ponce de León, former governor of Puerto Rico, believed that he found the fountain of youth, and science itself tells us that life first emerged from water.

Furthermore, each of us has our own stories and beliefs about the water and water products we embrace as being the best. From bubbling mountain spring beer water to hot spring spas and beyond, we are all crazy for water.

With that said, the absolute best starting point for us to educate ourselves about the physical properties of water and then to separate fact from fiction and folklore is to look at the nature of water.

We frequently refer to water verbally as H-two-oh, referring of course to the chemical representation of water, which is H2O. Many also know that this designation means that water consists of 2 atoms of hydrogen and one of oxygen. This jargon has been handed down between generations and we all speak fondly and comfortably of water in these terms.

If I can impart only one piece of information about water, I want it to be the very properties contained in this chapter, for once we understand these properties, the answers to things like contamination, personal water filters, pollution, treatment plants and infrastructure become much clearer.

The following diagram depicts what the water molecule looks like, and it also describes the behavior of water.

FIGURE 1

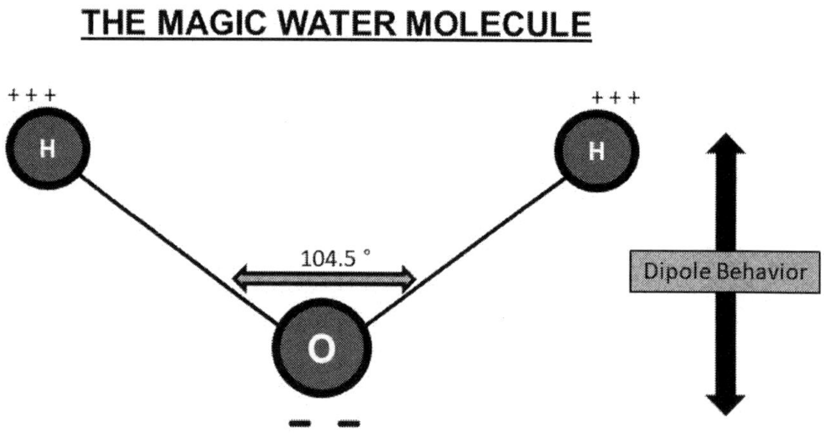

The first thing to notice about the water molecule is that it has a slight (partial) positive electrical charge at the parts near the hydrogen atoms, and a slight (partial) negative charge at the end where the oxygen atom is. In chemistry this is called a dipole moment. In our lives this has significant implications.[7]

The charged, dipole nature of the water molecule cause it to dissolve many other compounds and materials. For instance, the dipole charac-

7 https://www.worldofmolecules.com/solvents/water.htm.

teristic allows water to dissolve common table salt, NaCl, into sodium ions Na+, and chloride ions Cl-. This characteristic has caused scientists to label water as the "universal solvent," as it dissolves more substances than any other liquid known to humankind. In the next chapter, when we talk about water purity, we will see why the dipole nature of water is so crucial.

The bonds between the oxygen and hydrogen atoms are known as hydrogen bonds. Their extreme strength is responsible for the strong chemical stability of the water molecule. It also explains why the idea of an engine that runs on water remains confined to science-fiction conventions and school experiments.

Hydrogen is the most explosive element in nature so it seems natural that inventors would look to water as a source of energy. The green, inexhaustible supply certainly sounds attractive, doesn't it? And look, some inventors have created engines that run on water. So why are we not pouring everything into this technology?

It goes back to the strength of the hydrogen bonds I spoke about. They are so strong that the energy required to break them is far greater than the energy harnessed from breaking them. So, the next time you see one of these magical water engines, remember they consume more energy than they produce.

I will also mention that the angle between the three atoms of the water molecule is shown to be 104.5°. There are countless volumes of works that discuss the importance of this angle, and the debate about which water properties it is or is not responsible for has caused more than one scientific argument.

The angle is an identifier of this unique molecule and for a long time was a puzzle because simple geometry dictates that the angle should be 109.5°. To make a boring story a little longer, the molecule has a lone pair of electrons that force the angle to be a bit smaller. This is called minutiae, and you can learn more in the following magnificent footnote.[8]

Finally, water has some unique properties among other liquids. Let us look at them:

1) Water expands when it freezes. Other liquids shrink when they freeze.

2) Liquid water is slightly heavier than frozen water. For this reason, water freezes from the top down. Also, for this reason fish always have a place to swim when a lake freezes.

3) Water is a tremendous insulator compared to other liquids. It is an excellent liquid for cooling things. Car radiator anyone?

4) As mentioned above, water is the best solvent known to humankind. Nothing dissolves more substances.

5) Water has the highest surface tension of any liquid. This allows the surface of water to resist penetration by other materials. It explains why insects can float and even walk on the surface of water. It also explains why bubbles are round.

6) Water is the only liquid that when frozen is slippery and can be skated on. The fun part is that science still cannot explain why this is.

If these properties pique your interest, a good book on the chemistry of water might be in order. For now, we are just covering the basics.

---

8 https://www.chem1.com/acad/sci/aboutwater.html.

3.
Pure Water

Ask a hundred different people to give you an explanation of what pure water is and you will receive a hundred different explanations, along with a hundred personal choices for which water is the purest. Pick up a hundred different water products and you will witness a hundred different claims and testimonials about water purity. Furthermore, if you ask a hundred different scientists how to purify water, again, you will receive a hundred different answers.

Some people will be surprised at this next statement and some may even be angered; however, pure water does not exist anywhere in nature. Not only that, but it is impossible for pure water to exist anywhere in nature. It is also a fact that there is no place an average consumer can go to obtain 100% pure water.

Equally as startling is that even if you did have pure water in your possession, the very moment that you open the container it would no longer be pure water, but instead it would be contaminated water.

One of the little-known characteristics of water is that being contaminated is the happy and natural state of water. When you see a green pond, or yucky-looking dog water that was left outside for a few days, that is how water is supposed to look. Pure water is an unnatural condition for water and if you try to make pure water, it will fight you with a vengeance to contaminate itself.

Why is this? you ask. Go back and look at the water molecule drawing in chapter 2. It is the dipole characteristic, the slight electrical charges at each end, that prevents water from being naturally pure. Those electrical charges constantly seek neutrality. The way they find this neutrality is by combining with opposite charges. The negatively charged end of the molecule strives to find available contaminants bearing a positive charge, and the positively charged end of the molecule seeks to find contaminants with an available negative charge.

Water strives to be at peace, or what scientists call equilibrium. Water is stable only at equilibrium and it finds this resting state by dissolving and attaching to every available charged particle it encounters.

Let us look at a simple example of just how reactive water is.

If we were to take a beaker of pure water and let it sit, in a short time we would find that the pH of the water, which started neutral (pH 7), would soon settle at a pH of roughly 5.5. The reason for this is that the water would have absorbed CO2 from the surrounding atmosphere and formed carbonic acid. Even in the absence of contaminants in its container, water finds a way to degrade.

Furthermore, let's talk about what the pure water might be contained in. If we were to circulate pure water through our household pipes, we would find that in a short period of time the copper would be dissolved. Your water would be brown and poisonous and eventually the pure water would destroy your plumbing, fixtures, and would cause leaks throughout your home.

The closer to total purity water is, the more quickly it will dissolve things around it. Even at equilibrium, water will continue to dissolve most of what it encounters. The reason for this is that even though the water molecule has opposing charges at each end, those charges are more attracted to the electrical charges of contaminants than they are to each other. As such, the charges of the water molecule are always active, always looking for new material to dissolve.

Science states it this way: The charged poles of the water molecule are in constant flux. They can connect electrically to far more than just one available electrical charge. Picture the water molecule as a juggler, constantly touching objects but never holding on to any single one permanently. It is this property that has given water the title of "universal solvent."

When we speak of hard water, what we are discussing is water that has encountered calcium salts, generally in the form of limestone or chalk found in river, lake, or well beds. The water dissolves the calcium and voila, it winds up in your drinking water.

This illuminates why it is so important to prevent water from coming into contact with toxic materials, like lead, aluminum, and chromium, because water will dissolve those substances and deliver them in solution for us to drink. It also illuminates why it is so important for us to stop putting toxic material in our landfills, trash cans, drains and sewers. If we allow toxic material to enter our earth, water will dissolve them and bring them back to our tap.

Moving forward then, for us to properly address water problems, we now understand that any discussion making claims about pure water as it applies to a bottled product, water found in nature, water filtered by a consumer product, or even in a babbling mountain brook is a discussion based on errant information.

In the world of science, industry, and medicine, pure water can only be obtained through the engagement of highly engineered, complex technology. Surprisingly, it is only by way of even more complicated technology and greater difficulty that the water can be maintained as pure. There are applications where high-purity water is absolutely mandated, but in the world of drinking water it simply does not exist.

There are different ways to measure water purity and different industries and conventions describe water purity by different standards. Let us look a bit deeper into these descriptions.

From a mineral contaminant standpoint, there are three primary methods of measuring water purity. Conductivity, resistivity and total dissolved solids (TDS):

- **Conductivity or resistivity** measures respectively the ease or difficulty with which electricity passes through a given amount of water. Resistivity is used to convey very pure water, while conductivity is generally used to describe water that is less pure. Both measurements describe the ionic, or mineral content, of water. In science we call it a measurement of ions, dissolved solids, or electrolytes.

  Resistivity is expressed in ohms while conductivity is expressed in mhos. The reason for using two conventions as we will see, is that doing so enables the scientist to use numbers that make sense and are easy to digest at the further ends of water purity/contamination.[9]

  Think of it this way: If I told you that a shipping container of coffee beans weighed 3 tons, this would be relatively easy

9 https://chem.libretexts.org/Bookshelves/Inorganic_Chemistry/Map%3A_Inorganic_Chemistry_(Housecroft)/06%3A_Structures_and_energetics_of_metallic_and_ionic_solids/6.08%3A_Bonding_in_Metals_and_Semiconductors/6.8A%3A_Electrical_Conductivity_and_Resistivity.

to assimilate and get a mental picture of. However, if I told you the container weighed roughly 2,800,000,000,000 micrograms (2.8 trillion), you would be bewildered and driven to use a calculator to make sense of it all.

In a reverse fashion, if I told you to use a half ounce of salt in a recipe, you could easily comprehend this. However, if I told you to use 0.000015625 tons of salt, you would again be bewildered and require a calculator.

In both of my examples, both sets of numbers represent the same weight. The takeaway here is that at certain weights, one of the two measurements may be more practical and easier to understand.

Totally pure water exhibits an electrical resistance of 18.2 million ohms/cm-cm (18.2 megohms/cm-cm).[10] Essentially, electricity will not pass through this water. If you were standing in a pool of such water and some dastardly person stuck a live electrical wire into the water, you would be unphased. The current would never reach you.

For comparison, now let us represent this same ultrapure water in terms of conductivity instead. This conductivity is stated as 0.000018 micromhos. As you can see with high water purity it is easier to talk about 18.2 megohms than it is 0.000018 micromhos.

Conversely, if we wanted to describe water that had a lot of ionic contamination, such as a typical municipal tap water processed from a lake, we would find it to be about 400 micromho. If we instead wanted to represent that quality in terms of resistance, we would state that the water quality was 0.0022

10 http://www.myronl.com/PDF/application_bulletins/di_ab.pdf.

megohm. Again, it is easier for us to comprehend and communicate larger, whole numbers, so for this water, micromhos would be a better scale to speak of.

In assessing ionic purity of water, a little trick to gaining a working grasp of these two measurement scales is to always remember that conductivity and resistivity are reciprocals of each other. 1 megohm = 1 micromho. Also 2 megohm = 0.5 micromho. Hence the convention of mho, it is ohm spelled backward.

The third way of measuring water purity from an ionic/metals/electrolyte standpoint is to describe how many ions are in the water. This convention is called total dissolved solids and is expressed in measurements of milligrams per liter (mg/l) or parts per million (ppm). For an overall understanding of this measurement convention mg/l = ppm.

The following abbreviated chart gives a visual example of how conductivity, resistivity and TDS relate to each other.

TABLE 2

TDS (ppm - mg/l)	RESISTANCE	CONDUCTIVITY
0.0352 ppm	18 megohm/cm^2	0.055 micromho
0.06 ppm	10 megohm/cm^2	0.1 micromho
0.6 ppm	1 megohm/cm^2	1.0 micromho
1.2 ppm	0.5 megohm/cm^2	2.0 micromho
6.0 ppm	0.1 megohm/cm^2	10 micromho

- **Organic purity** – Aside from mineral purity, it is important to know how pure water is organically. In science, when we speak of organics, we are referring to chemicals, particles and contaminants of any kind that have the carbon atom contained in them.

  Examples of organic contaminants are sugars, gasoline, methane, antibiotics, plastics, bacteria, viruses, plant/vegetation, algae, fungi, alcohols, esters, ethers, and countless other substances. Some organic contaminants will affect the conductivity/resistivity reading, while others will not.

  Testing for the presence of organics and identifying them can be quite simple or also exceedingly difficult depending on the contaminant. Certainly, we could detect diesel fuel by smell and vision, but many carcinogenic are dangerous in such small amounts that we can drink them and never know it.

  What also makes organics so dangerous is that there are not any in-line meters that can constantly measure them. Only highly specific and sensitive analytic equipment can detect such organics. Because of this it is crucial that water providers sample multiple points of use and perform frequent testing. This sneaky nature of organics makes delivering pure water a technological challenge.

  While we are talking about pure water, you might find it interesting to consider a few standards that have been put in place to determine water purity. Let us look at ionic purity, in other words conductivity/resistivity. Let's list water in order from low purity to high purity.

TABLE 3

Less Pure	Distilled water – 0.2 megohm
	Reverse osmosis water – 0.2 megohm – 1.0 megohm depending on feed water quality
	USP water for injection (WFI) – 0.8 megohm (IV bag, injectable drugs)[11]
	Spot-free car wash rinse – as pure as 1 megohm or better
	Water for rinsing electronic parts – 10 megohm
More Pure	Power plant steam production – 18.2 megohm or close

Some might be surprised to learn that most industrial water applications require water that is far ionically purer than water that goes into an IV bag or an injectable drug. Can you imagine that water required for spot-free automobile rinsing can be purer than water in your hospital IV bag?

Now before you go down to the Spiffy Wash and start filling up your IV bag, let's harken back to the earlier part of this chapter where we learned that there are different gauges to determine water purity. The WFI does not have a strict requirement on ionic material, other than it is the ions are safe, but from a sanitary perspective, regarding things like bacteria, viruses, and chemicals, it has very strict standards that the car wash water could never meet.

11 https://www.gmp-compliance.org/gmp-news/overview-of-the-requirements-on-wfi-in-the-eu-usa-and-japan.

Here is what I wish for you to take away from this chapter:

- Truly pure water is not offered commercially to consumers.
- There are different standards to describe purity, and acceptable purity for one application may not meet the specifications for another.
- Pure water does not exist in nature.
- Pure water is extremely reactive and corrosive.
- The natural state of water is a contaminated one.

4.
Water Contaminants

So, if contamination is the natural state of water, does that mean we are all doomed to drink bad water? Of course, not! However, as we prepare to defeat the impending water apocalypse, it is important that we become educated about water contaminants. We need to understand which contaminants are harmless and which ones are dangerous. Furthermore, with an understanding of contaminants, we can expand our water skills to where we can truly interpret a water analysis. That in turn will enable us to make better choices about which technologies remove which contaminants. At the end of the day, grasping the science of water contaminants will allow us to make safe, healthy decisions about the water we drink.

Chemistry can be dull and cumbersome, and this chapter may strike some as a chemistry textbook. However, it is meant to serve as a reference chapter, one that will have more use and meaning as you explore different and new areas of water. With that warning, let us get into the science of water contaminants.

There are only two fundamental types of water contamination, and each of those types has two different classifications. Water contaminants must be either organic contaminants or inorganic contaminants. In chemistry, the word *organic* signifies the carbon molecule. To be slightly more specific, organic chemistry deals with molecules in which there are also carbon-hydrogen bonds within the compound. Conversely, the word *inorganic* signifies the absence of carbon.

We do not want to confuse the word *organic* in the science world, with the common use of the word *organic* in the food world. Scientifically, all food is organic. Even artificial flavors and colors, even pesticides and preservatives are organic. Plastic, polyester, sugars, oil, grease, steel, and even gasses like methane and propane are all organic substances.

Continuing, take note that within these two groups of contaminants (organic and inorganic) are also two subclassifications of contaminants. All contaminants can be either dissolved or undissolved.

Dissolved contaminants are incorporated into the water and are said to be *in solution.* Undissolved contaminants will not go into solution, and even after vigorous stirring or shaking, will visibly either float around in the water or come to rest on the bottom of the container.

The following table will help us to identify and classify contaminants as we develop our water toolbox.

Classification of Water Contaminants

TABLE 4

INORGANIC CONTAMINANTS	ORGANIC CONTAMINANTS
· Dissolved · Undissolved	· Dissolved · Undissolved

Let us examine each type of water contaminant separately.

Dissolved inorganics – The most common dissolved inorganic substances are the minerals/salts that we see leaving spots on our dishes, cars, and windows. Sodium, calcium, iron, chlorides, sulfates, magnesium, manganese, and potassium are common to this group. These elements are found in the rocks, dirt, sand, and clay that our water comes into contact with. These types of contaminants are also found in all drinking waters and are among the friendly variety of inorganic contaminants. Some of them are even called minerals. While not generally harmful they can affect the taste of the water.

Other inorganic contaminants, which are known in chemistry as metals, are also often found in our water. These metals include selenium, iron, barium, lead, chromium, nickel, silver, and copper among many others.

In low concentrations, many of these metals are cherished as healthful minerals. In greater concentrations, however, these types of trace elements can be toxic. Even in concentrations as little as 1 part per million some of these contaminants can cause devastating health problems and even death.

In addition to single-element dissolved inorganics there are also soluble salt compounds to consider. The rules and exceptions to which inorganics are soluble versus insoluble would fill a textbook, but as a rule, most chlorates, perchlorates, nitrates, acetates, sulfides, chlorides, bromides, fluorides, and iodides, as well as ammonium salts are commonly soluble. This means that when put into water they dissolve (disassociate) into both positively charged and negatively charged particles, again, commonly called ions.

In general, in any water sample there must be an equal component of negatively charged particles to positively charged particles, with most inorganic contaminants in water being dissolved salts, such as sodium chloride (Na+ Cl-).

If I can impress nothing else upon the reader with regard to dissolved contaminants, it is to always remember from chapter 1, that water, as precious and necessary as it is to our lives, is an aggressive, active solvent that can do us as much harm as it does good.

It is crucial that we understand that the powerful solvent action of water, depending on its surrounding environment, can change drinking water quality from safe to toxic in the blink of an eye. If the salts our water meets are safe, like sodium chloride, then we do not have a concern; however, if the salts are toxic, then the water dissolving the salts will also become toxic.

Undissolved inorganics – These are also called insoluble inorganics. These materials contain no carbon-hydrogen bonds. They will not go into solution no matter how much the water is shaken. Examples include carbonates (limestone), oxides (rust), and most chromates, phosphates, silicates, silver salts, and sulfides.

In general, undissolved contaminants tend to build up and foul filtration equipment, pipes, and tanks. As such they are a big problem for treatment plants and industry.

An interesting point to remember is that a substance can be both a dissolved and an undissolved water contaminant. Depending on a variety of factors such as pH, temperature, and specific overall water chemistry, an element can present itself in different forms.

Iron for example can be dissolved in water, and many of you have tasted how nasty this contaminant can be. However, when oxidized with chlorine or high pH, the iron precipitates out as rust, and can then be filtered out.

It is important to know that even though a contaminant is undissolved in our water, we can still suffer great health problems from ingesting even small amounts of many of these chemicals. Because the pH of the normal human stomach ranges from 1.5 to 3, contaminants insoluble in the normal pH range of drinking water can become very soluble and toxic in the acid environment of the stomach. Once solubilized in the stomach, they can be freely absorbed into our blood and organs.

Always remember, it is not only the state of water that constantly changes, but the water-bound contaminants as well. The tragedy in Flint, Michigan, is a particularly good example of this. Under the chemical conditions of the original water supply, the lead in the city's pipes was not a contamination issue. However, once the chemistry of the city water changed, then it dissolved the lead, and the water became poisoned.

Following and documenting contaminants is difficult enough, but when we realize that contaminant toxicity will change with differing conditions, we get a glimpse of not only how important it is for treatment professionals to monitor and understand contaminants but also how important it is for us to begin understanding water contaminants is.

Dissolved organic contaminants – As a rule, complex organic compounds are not soluble in water. The chemical bonds in organic compounds are not polar in nature. In other words, they do not have positive or negative charges readily available to interact with the electrically charged ends of the water molecule. It is this electrical interaction that makes things water-soluble. It is the lack of this interaction that makes contaminants insoluble.

Some organic compounds such as sugars do dissolve in water, as do some lighter oils. In general, if an organic compound has 4 or less carbon atoms it will have a higher probability of being water-soluble. Also, smaller organic compounds that have electrically charged groups attached to them may also be water-soluble. Ethyl alcohol (C2H5-

OH) is one good example of such a compound. In the representative model seen below the hydroxyl group shown in brackets has a slightly negative charge making the molecule subject to being soluble in water.

Ethyl Alcohol

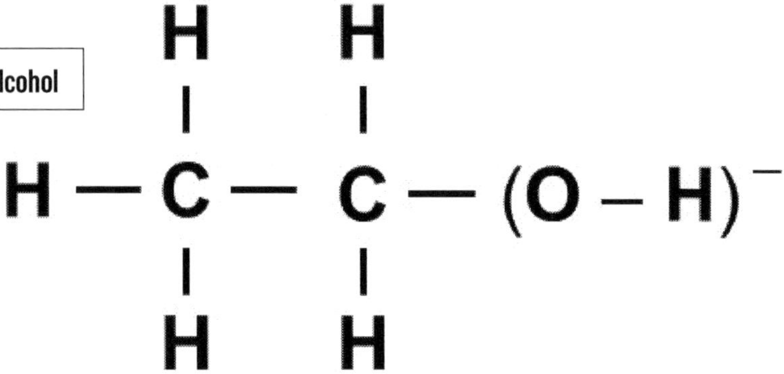

Industry is a major producer of dissolved organic compounds. It is safe to say that industry does a good job of keeping these contaminants out of our water supplies. Still, if these chemicals are put into landfill, they can eventually work their way into our drinking water. Many of them are toxic and we need to keep track of them.

The compounds that represent the highest probability of being water-soluble include sugars, alcohols, amines, acids, esters, and aldehydes.

You might find it interesting to learn that the biggest culprit in placing dissolved organics into our water, however, is Mother Nature. Bacteria, viruses, plankton, fungi, and algae continually feed on decaying vegetation. The metabolism of these living organisms creates compounds such as fulvic and humeric acids, proteins, amino acids, tannins and even vitamins. The next time you notice the color and odor of pond, lake, stream, or river water, you will be experiencing nature's bounty of dissolved organics.

Undissolved organic contaminants – So here we go! Now to the finale of water contaminants. To qualify for this group, a contaminant must contain the carbon-hydrogen bond and refuse to go into solution. Some of the most feared and nefarious water contaminants fall into this category. When these types of contaminants get into our water it makes the headlines, often damaging the environment, wildlife, and humans.

Because most of these compounds are complex hydrocarbons, they are not easily neutralized or removed from the environment. They include oil, grease, solvents, paints, and flammable methane in drinking water often associated with fracking operations. Both natural and synthetic pesticides fall into this category. Other contaminants in this category are benzene, phenol, and alcohols with more than 4 carbon atoms.

An additional class of problematic undissolved organic is known as trihalomethanes, or THMs. THM compounds are commonly identified as precursors of cancer. They are well monitored but commonly found in our drinking water. An alarming new discovery is that we may absorb more THM contaminants via hot showers than we do by drinking them.[12]

You may be surprised or even outraged to learn that this contaminant winds up in our drinking water as a direct result of municipal water treatment—that by and large, THMs wind up many of our drinking waters because of municipal water treatment regimens.

THMs and many other contaminants in this class are volatile organics, which means in time they will evaporate or can be boiled off. A granular activated carbon filter will also remove a lot of the THM content from drinking water.

The good news is that most municipalities do a stellar job of keeping THMs and all the contaminants in this category well below maximum safety EPA limits. We will talk more about THMs in chapter 5.

12 https://www.ncbi.nlm.nih.gov/pmc/articles/PMC1257669/#:~:text=Trihalomethanes%20(THMs)%20are%20the%20result,cancer%20and%20adverse%20reproductive%20outcomes.

Finally, organisms such as bacteria, viruses, mold, algae and fungi are also considered undissolved organic contaminants. The addition of chlorine is vital to keeping these contaminants to virtually undetectable limits. Again, our municipalities are very much in tune with keeping us safe from such toxic material.

There are two important takeaways from this chapter that I hope you now carry:

- Providing safe, secure water requires that we identify exactly which contaminants are present.
- Water contaminants come to us both from the activities of civilization and from Mother Nature herself.

Bonus contaminant: type #5 – So that my scientific colleagues do not get their underpants in a knot, there is another type of contaminant known as a colloid.

Definitions abound but essentially colloids are substances that stay suspended in other substances. Gels, milk, aerosols, dirty brown water, plaster, gelatin, and whipped cream are all examples.

The most common water contaminant in this category is colloidal silica. This is essentially microparticles of sand that are forever suspended in some water sources. They are so small that the forces of gravity will not pull them to the bottom of a container, as is the case for instance with dirt or rust.[13] What makes colloidal silica a miserable contaminant is that it readily plugs up filtration equipment resulting in increased labor costs and expensive plant shutdowns. In general, colloids are not something that are a big concern to you and me when it comes to our drinking water.

• • ● • •

13 https://colloidalsilica.nouryon.com/colloidal-silica-facts-what-is-silica/.

A question that I am often asked is, "What is the best water filter?" Now that we have illuminated just how many different types of contaminants there are, I trust that you are beginning to see that choosing the "best" water filter depends on what contaminants are in the water. The best filter for someone drinking Lake Erie tap water may not be the best filter for someone drinking tap water from a well in Arizona.

In the next chapter we are going to examine what the various filtration technologies are and look at what exactly goes into designing a proper water purification system.

5.
How to Read a Water Report

Now that we understand water chemistry a bit better, purity and the mechanism by which water becomes contaminated, let us take a look at a typical water report and learn how to make sense of it.

Different communities provide different levels of water reports. As always, the more information we consumers have, the better prepared we will be to make important decisions about the water we drink.

The ideal water report will be detailed with analytical results about both harmful and harmless/beneficial contaminants. It should share bacteria, volatile organics, dissolved organics, and inorganics. A good report will also share some of the noncritical chemical properties of water such as hardness, pH, color, and alkalinity.

One of the best examples of a water report you will ever see is put out by Memphis Light Gas and Water in Memphis, Tennessee.[14] The folks in El Paso, Texas, also do an amazing job of presenting a thorough

14 http://www.mlgw.com/images/content/files/pdf/Water%20Quality%20Table%20 2019_web.pdf.

water document to the public.[15] The following water report is an example of what a proper report should present. It was taken from the lost city of Aquatropolis, somewhere in the South Pacific.

To prepare us for reading an actual water report, let us start by examining the commonly used terms/abbreviations that allow us to make sense of a water analysis.

AL – Action level – This is the concentration level that if exceeded initiates action to be taken to remedy. The action may be a smile treatment adjustment, a boil order, or even additional equipment to be used. Frequent episodes of exceeding these action levels could indicate critical problems with a particular water source.

ND – Non-detectable – This means that the contaminant in question was below the ability of the test to detect. It does not mean that this contaminant is not present in the water. A reading of ND is reassuring to the consumer as it does designate that the substance is far below the maximum level for drinking water safety.

MCL – Maximum contaminant level – This is the level of contamination the EPA considers safe to drink. A value above this level is considered to be unsafe to drink and an action plan to correct it must be enacted.

MCLG – Maximum contaminant level goal – The level of a contaminant in drinking water below which there is no documented health risk.

Mg/l or ppm – Milligrams per liter and parts per million are the two scientific conventions most widely used to convey how much of a contaminant is found in water. Different contaminants have vastly different safe drinking levels. While 100 ppm of calcium is safe, 100 ppm of aluminum is fatally toxic.

---

15 https://www.epwater.org/our_water/water_quality/water_quality_faq.

µg/l or ppb – Micrograms per liter and parts per billion are the scientific conventions used to express contaminant levels that are toxic in almost non-detectable levels. 1,000 ppb = 1 ppm. 1,000 µg/l = 1 mg/l.

µmho/cm – Micromhos per centimeter – This is the measurement of electrical conductivity, of how easily electricity moves through water. Higher levels mean electricity passes through easier. The more salts, or ions, in the water, the more conductive it is. Conductivity readings can be converted to total dissolved solids (ppm of ions) by multiplying by 0.60. Most city water sources in the US fall in the 100–300 micromho range, which is 60–180 ppm total dissolved solids. There is no real safe level with this reading other than the EPA likes to see a total TDS level of 500 ppm or less, which would translate to a conductivity of 834 µmho. Perrier® water once advertised 505 ppm, above the EPA action limit, but was not harmful at all since the concentrations were all healthy minerals. Conductivity is a good indication of the general stability of a water source and is a valuable tool to help analyze water treatment.

NTU – Nephelometric turbidity units – This is a measurement to describe how much color and clarity there is in a water sample.

KEY WATER DROP – While the following water report hails from a fictitious town, the maximum allowable concentrations shown in the first column are the EPA Safe Drinking Water Act limits for those contaminants.[16] If you ever need to know those maximum concentrations and have this book at hand, come revisit chapter 5.

16 https://www.epa.gov/ground-water-and-drinking-water/national-primary-drinking-water-regulations.

CITY OF AQUATROPOLIS ANNUAL WATER REPORT

TABLE 5

CONTAMINANT/ TEST	MAXIMUM ALLOWABLE	LAKE 1	LAKE 2	AVERAGE	ACTION REQUIRED IF ANY
Clarity/ turbidity	2.0	0.18	0.19	0.185	No action required

- This result is very typical for most drinking water supplies found throughout the US. These values are deemed to be excellent.
- *Caused by both dissolved and undissolved organics.*

TABLE 6

MICROBIOLOGICAL					
Total coliform	Positive	-0-	-0-	-0-	No bacterial coliform
Fecal coliform	Positive	-0-	+	-0-	Boil alert, chlorinate, flush tank and lines, retest

- Fecal bacteria were found in Lake #2. The city issued a boiling alert, re-chlorinated the system, flushed, and then retested to eliminate the contamination. Normally drinking water should not contain coliform bacteria. When it does, boiling orders are the norm.
- *These are undissolved organics.*

TABLE 7

ORGANIC CHEMICALS PESTICIDES (mg/l)					
ALACHLOR	0.002	ND	ND	ND	No action required
ATRAZINE	0.003	ND	ND	ND	No action required
CHLORDANE	0.002	ND	ND	ND	No action required
ENDRIN	0.002	ND	ND	ND	No action required
GLYPHOSATE	0.002	ND	ND	ND	No action required
HEPTACHLOR	0.0004	ND	ND	ND	No action required
HEPTACHLOR EPOXIDE	0.0002	ND	ND	ND	No action required
LINDANE	0.0002	ND	ND	ND	No action required
METHOXYCHLOR	0.04	ND	ND	ND	No action required
POLYCHLORINATED BIPHENYLS (PCBs)	0.0005	ND	ND	ND	No action required
SIMAZINE	0.004	ND	ND	ND	No action required
TOXAPHENE	0.003	ND	ND	ND	No action required

- Based on geography the actual pesticides tested for may vary. Different crops and insect populations across the nation often require different chemistries. This phenomenon mandates that we have local specificity in our different treatment plants. An analysis that provides safety for one city may neglect toxic contaminants for another.

- These are dissolved organics.

TABLE 8

SEMI-VOLATILE ORGANIC COMPOUNDS (mg/l)					
BENZO(a)-PYRENE	0.0002	ND	ND	ND	No action required
DI(2-ETHYLHEXYL) ADIPATE	0.4	ND	ND	ND	No action required
DI(2-ETHYLHEXYL) PHTHALATE	0.006	ND	ND	ND	No action required
HEXACHLOROBENZENE	0.001	ND	ND	ND	No action required
HCL-CYCLOPENTADIENE	0.05	ND	ND	ND	No action required

- Semi-volatile compounds (SVOCs) are a special type of volatile organic compound put into a separate category, characterized by higher molecular weights and higher boiling points than the standard group of volatiles listed in the next section. Some refer to them as polycyclic-aromatic hydrocarbons or PAH compounds.

- SVOCs are toxic and dangerous because they vaporize at room temperatures and the gases released highly toxic. SVOCs are found in many household products so it is crucial that they not find their way into our drinking water.

- SVOCs are newer toxins that our plants were not even on our screen a few decades ago. Treatment plants are being challenged more than ever before with new, never-before-tested contaminants.

- These are classified as dissolved organics.

TABLE 9

VOLATILE ORGANIC COMPOUNDS (mg/l)					
BENZENE	0.005	ND	ND	ND	No action required
CARBON TETRACHLORIDE	0.005	ND	ND	ND	No action required
1,2-DICHLOROBENZENE	0.6	ND	ND	ND	No action required
1,4-DICHLOROBENZENE	0.075	ND	ND	ND	No action required
1,2-DICHLOROETHANE	0.005	ND	ND	ND	No action required
1,1-DICHLOROETHYLENE	0.007	ND	ND	ND	No action required
CIS-1,2-DICHLOROETHYLENE	0.07	ND	ND	ND	No action required
TRANS-1,2-DICHLOROETHYLENE	0.1	ND	ND	ND	No action required
DICHLOROMETHANE	0.005	ND	ND	ND	No action required
1,2-DICHLOROPROPANE	0.005	ND	ND	ND	No action required
ETHYLBENZENE	0.7	ND	ND	ND	No action required
MONOCHLOROBENZENE	0.1	ND	ND	ND	No action required
STYRENE	0.1	ND	ND	ND	No action required
TETRACHLOROETHYLENE	0.005	ND	ND	ND	No action required
TOLUENE	1.0	ND	ND	ND	No action required

1,1,1-TRICHLOROETHANE	0.2	ND	ND	ND	No action required
1,1,2-TRICHLOROETHANE	0.005	ND	ND	ND	No action required
TRICHLOROETHYLENE	0.005	ND	ND	ND	No action required
1,2,4-TRICHLOROBENZENE	0.07	ND	ND	ND	No action required
VINYL CHLORIDE	0.002	ND	ND	ND	No action required
TOTAL XYLENES	10.0	ND	ND	ND	No action required
TOTAL TRIHALOMETHANES	0.080	0.007	0.014		No action required

- Volatile organics are classified as a dissolved organic, see chapter 4. They come from many sources and include things like solvents, gas, diesel fuel and more. The good news is that most of these compounds are easily removed by our municipal water treatment plants.

- As we mentioned in the previous chapter, one of the most talked about volatile organic contaminants, a suspected carcinogenic, trihalomethanes are also known as THMs. The actual safe drinking limit for THMs is listed in this section.

Like it or not, chlorine, even though it is in many ways a double-edged sword, remains the most effective, safe, and cost-efficient way to keep bacteria from growing in our drinking water. Remember, most of the worst plagues to ever attack humankind have come from non-chlorinated water. Typhoid fever quickly comes to mind!

Unfortunately, when chlorine is added to methane gas, THM can form. In water, the methane comes from decomposing vegetable and animal matter. Leaves, pine needles, food waste, dead animals, algae, plankton and more serve as food for bacteria. As bacteria digest and metabolize, methane gas is given off. Boom, a dangerous but unavoidable marriage.

Surface water sources (lakes and rivers) have a higher probability of forming THMs as opposed to groundwater sources. Surface waters, lakes, ponds, rivers, and streams are a playground for both animals and plants, thereby providing higher levels of methane than water that comes from deep within the earth.

Again, the good news is that municipal water systems do a great job at keeping THMs well within safe drinking water levels.

- These contaminants are dissolved organics.

TABLE 10

INORGANIC CHEMICALS (mg/l)					
ALUMINUM	0.2	0.06	0.08	0.07	No action required
ANTIMONY	0.006	ND	ND	ND	No action required
ARSENIC	0.01	ND	0.001	ND	No action required
BARIUM	2.0	0.028	0.032	0.030	No action required
BERYLLIUM	0.004	ND	ND	ND	No action required
CADMIUM	0.005	ND	ND	ND	No action required
CHROMIUM	0.1	0.008	0.006	0.007	No action required
COPPER	1.3	0.004	0.12	0.062	No action required
LEAD	0.015	ND	ND	ND	No action required

MAGNESIUM	Not Listed	4.2	2.4	3.3	No action required
MANGANESE	0.05	0.003	0.005	0.004	No action required
MERCURY	0.002	ND	ND	ND	No action required
MOLYBDENUM	NS	ND	ND	ND	No action required
NICKEL	0.1	ND	ND	ND	No action required
POTASSIUM	NS	0.68	0.24	0.46	No action required
SILVER	0.1	ND	ND	ND	No action required
SELENIUM	0.05	0.006	ND	0.003	No action required
THALLIUM	0.002	0.0008	0.001	0.0009	No action required
ZINC	5.00	ND	ND	ND	No action required

- Inorganic contaminants are classified in chapter 4 as dissolved inorganics. These contaminants vary greatly across the globe and are affected by both the soil that contains/holds the water, and of course like all other contaminants by what is put into the water by local manufacturing and citizens. All water will have inorganic contaminants. Some of them are tolerable in high concentrations while others are fatal at the smallest of concentrations. As always, each contaminant must be considered individually.
- These contaminants are dissolved inorganics.

TABLE 11

OTHER CHEMICALS (mg/l)					
CHLORIDE (mg/L)	250	9.0	8.2	8.6	No action required
COLOR (units - PCS)**	15	<6	<6	<6	No action required
CYANIDE (mg/L)**	0.2	ND	ND	ND	No action required
DETERGENTS - MBAS (mg/L)**	0.5	ND	ND	ND	No action required
FLUORIDE (mg/L)	4.0	0.8	0.8	0.8	No action required
IRON (mg/L)	0.3	0.004	0.008	0.006	No action required
NITRATE (as Nitrogen) (mg/L)	10.0	0.08	2.2	1.5	No action required
NITRITE (as Nitrogen) (mg/L)	1.0	ND	ND	ND	No action required
ODOR (TON)**	1.0	1.0	1.0	1.0	No action required
pH (units)	6.5-8.5	7.8	7.6	7.7	No action required
SODIUM (mg/L)**	NS	22.5	20.0	21.25	No action required
CONDUCTANCE (umho/cm @ 25°C)	900	200	190	195	No action required
SULFATE (mg/L)	250	11.0	26.0	18.5	No action required
TOTAL DISSOLVED SOLIDS (mg/L)**	500	128	120	124	No action required

- This group of contaminants was placed in a group that describes more of the overall physical/chemical nature of the water as opposed to singling out hazardous/toxic contaminants. Some contaminants like sodium and magnesium in the table before are not even listed as contaminants because the body has a tolerance level far above what has ever been found in water. In general, it makes no scientific sense to test for toxicity at levels never found in nature. This group of contaminants, while important and containing some toxic materials, are just not generally found to be problematic. They do, however, yield a lot of information about surrounding geography, physical condition of the bodies of water. Conductivity and TDS for instance are two of the most stable components of a water analysis, often not varying by more than a few percent over a century. Sudden changes in these parameters are rare and would indicate a major catastrophic event.

- pH is an exception to this casual look in this contamination group. pH is often controlled by the addition of alkaline or acid chemistry to achieve proper balance. When the pH of water is too low, less than 6.5 it is considered acidic and the water will leach harmful metals such as copper, zinc, nickel and lead into the drinking water. This is one parameter that must be watched closely.

- *These contaminants represent all four types of contaminants.*

TABLE 12

ADDITIONAL MONITORING PARAMETERS (mg/l)					
ALKALINITY as CaCO3(mg/L)	44	56.2	80.6	68.4	
CALCIUM (mg/L)	8.8	14.8	22.8	18.8	

HARDNESS as CaCO3 (mg/L)	NS	70	50	60	
HARDNESS (grains/gal)	NS	12	12.6	12.3	
PHOSPHATE (mg/L)	NS	4.0	12.0	8.0	
TEMPERATURE (°F)	NS	48.3	44.3	46.3	
TOTAL ORGANIC CARBON	NS	12	12	12.0	

- This last group of parameters is generally not included in a typical consumer water analysis report but should be. Nothing on this last table will mean much to the average person, but these are parameters that allow a water engineer to perform calculations about water safety that go beyond the typical contaminant concentration level.

For instance, there is a calculation called the Langelier Saturation Index (LSI) that in essence measures the tendency of water to either cause scale or cause corrosion. This is a mathematical calculation, and it involves using some of the parameters in these non-contaminant-specific tables. The LSA calculation utilizes pH, TDS, calcium and alkalinity. A positive LSI means that the water tends to leave scale. The higher the number, the higher the tendency. A negative LSI indicates that the water is corrosive. The lower the number, the higher the corrosion tendency. In the Flint, Michigan, tragedy the water became corrosive and caused lead in the pipes to be dissolved into the drinking water. Had a technician performed a proper LSI they would have discovered this problem and stopped it before poison was delivered to the residents.

The LSI is not the only water test or calculation that is normally not included in a water analysis but probably should be. Other tests only apply to certain industrial or agricultural uses. One such test is called a Silt Density Index (SDI), which tells the engineer what the tendency or water is to foul/plug a reverse osmosis system.

There are a couple of very important pieces of knowledge that I want you to take from this chapter.

1. In the introduction I mentioned that one of the problems facing us with regard to water is that it is an extremely complicated substance. As you look through the water analysis example I have provided, if you are like me, the sheer number of different contaminants that we must consider is mind-boggling. Which ones are harmful? Which ones are safe? At what concentration does a harmless contaminant become harmful? The next time you get mad at the local water authority, put yourself in their shoes because accounting for all these, and new or hidden poisons is not easy.

2. Whenever there is a water problem the average citizen asks about testing. Everyone wants their water tested. I am sure many of you have a home water softener or other filter, maybe a well system in which the local water expert comes out and tests your water. Usually, they have a handheld meter and perhaps a couple of test tube-type procedures that measure certain water contaminants by the color left in the test tube. The technician visits, changes your filters, runs a few tests and tells you that your water is fine. Really?

   Now, I am not bashing your local water service rep, but at the same time understand that to do a thorough water analysis like the one in this chapter it requires a lot of expensive equipment. Equipment that includes, but is not limited to, high-performance liquid chromatography, mass spectroscopy, pH meters, conductivity meters, colorimetric kits, an organics furnace, autoclave, pure water system, incubator and much more. A proper water lab could easily cost $1 million or more.

Add to this that these tests take time and must be run individually. We as consumers want instant satisfaction. We don't want to hear about technology and science, and lab procedure, damn it, just tell us what the one test is we can do to tell us if our water is safe. As you can see, there is no one test. Proving beyond a shadow of a doubt that water is 100% safe to drink takes many painstaking steps and it is expensive.

To make matters worse, if you remember earlier, I said that water is unstable. Water is never at rest and it is always looking to dissolve new contaminants?

Well get this, even if we have a freshly prepared, thorough water analysis like the one in this chapter, it is no guarantee that our water is safe. Why? It is because the chemistry of water is always changing. The water analysis we have in front of us is a statement about the past. It is a report that says on the day this sample was taken, it was safe. The test tells us nothing about what the water analysis looks like after it goes traveling down the distribution system, water tower and then to your home.

The good news is that by doing repeated tests and multiple sample points along the system, a municipality can determine just how stable their water supply and treatment process is, and some exceptionally reliable predictions about what you and I are drinking at the end of the pipe.

Also good news is that in most places, the bulk of the contamination levels don't sway out of range very often. However, we must realize that some contaminants are harmful in concentrations as little as a few parts per billion, so for some contaminants it does not take much to make our water toxic.

Whether we like it or not, providing safe water to consumers will always require the municipality to make a lot of generalizations that today's water should be remarkably like yesterday's water. Still, no matter how good the testing and treatment protocols are, you and I are always guinea pigs to some degree when we open the faucet and take a drink.

Later in the book we will talk about which technologies, policies, and procedures we can enact in our own homes to ensure a higher level of personal water security.

3. Perhaps most importantly, if you are concerned about your drinking water, simply download your local water analysis, and armed with what you learn in this book, dissect it. I encourage you to call, and even visit your local water authority. If you have water questions, they will help you. Your local water department is a great source of water education.

   We can rest assured that nowhere in the world do municipalities do a better job of providing safe drinking water than they do right here in the United States of America.

   This is not to say, however, that the changing world we live in is posing ever-increasing water threats to us. Let's learn more and be vigilant.

As we move forward, chapters 6–12 comprise what I like to think of as our basic water doctor toolbox. These chapters are meant to serve as a resource you can come back to as you consider new products, claims or even need to answer technical questions.

As we move forward, we will apply these tools to water issues that we face every day.

- What is in bottled water and which one is the best?
- What is the best home water filter?
- What is the purpose of a boiling order?
- Is distilled water pure?
- Water for our radiator, CPAP, pets, cars, and fish tanks.
- Climate change, conservation, and drought.

6.

Water Purification Technologies – Reverse Osmosis

People always want to know: "What is the best water filter?" or "What is the best water filtration technology?" In solving our own contamination problems, it is normal to just want a solution. We want a simple answer that guarantees us safe, secure water.

Unfortunately, when it comes to water there never is a best. Each water sample has different chemistry and physical characteristics and what works on one water source may not work on another water source. If we want the best system then we must realize that finding the best must always begin with what we covered in chapter 5, the water analysis.

With a current water analysis in hand, we can now go about finding or designing the best water system for us. In addition to our water analysis, we also now understand and can recognize which of the four types of water contaminants we are facing.

The next step then is to analyze, pick, choose and assemble exactly which technologies we need to achieve the safe, secure water we desire. What follows in the next few chapters is an identification and explanation of what the major water purification technologies are, what their strengths and weaknesses are, and how they fit together in an overall water purification system.

At the end of these technology chapters, we will examine a comparison chart of the various technologies to determine which technologies remove which contaminants. This will be a great tool and quick reference in your own water system considerations.

Before we go any further, however, I would like to address the widely used terminology of *water filter*. This is an overused and misinformed generality because not every water purification or water treatment technology is a filter. Technically, a filter is a screen, it has pores in it that allow particles of a certain size to pass through it, while larger particles are screened out. From this point forward I would like you to use proper terminology for water treatment/water purification technologies. So, for instance, a disposable sediment cartridge is a filter but a water softener is not.

Now, let's get on to the business of learning about these different technologies. The first technology we want to look at is reverse osmosis, also called RO.

RO can safely be called the world champ of water filtration technologies. By itself, it will remove all four categories of water contaminants and make exceptionally clean water out of virtually any type of feed water. Does this mean the RO is the best? Is it a panacea? Is the one-stop product that is all we need to know about water? Unfortunately, no. Once again, as you will hear me say a thousand times in this book, the effectiveness of any technology depends on the nature of incoming water (the water analysis), and factors like daily volume required, available pressure, flow rate, electrical service, space, and other factors. RO, just like every other water purification technology, can be a dream and it can also be a nightmare.

To understand reverse osmosis, we must first understand osmosis. Osmosis is a scientific principle that we can find in action throughout nature. Osmosis is the mechanism by which nutrients and water move freely back and forth across both plant and animal cell walls. When we consider the chemical concentrations on either side of the cell wall, we can predict the movement of particles. This principle is invaluable in studying things like nutrition, hydration and even the effectiveness of medicines. In water, the principle has been developed and refined into a wonderful technology that literally over and over has proven to be lifesaving.

The foundation of osmosis is something called a semipermeable membrane. A semipermeable RO membrane allows water to pass freely through it in both directions; however, the pore size of the membrane is so small that contaminants such as bacteria, calcium, pesticides and more cannot pass. As a result, in the operation of the RO, they are said to be rejected. The layperson would say that these contaminants were filtered. However, since you are now all well on your way to becoming water doctors, you should get accustomed to the word *rejected.*

Figure 2 demonstrates the principle of osmosis. In this experiment we create a container of water that has two equal chambers separated by a semipermeable membrane. Again, the nomenclature *semipermeable* means that some substances will pass freely back and forth through the membrane, while others will be screened, or rejected.

In this first figure, on each side of the membrane we pour in identical levels of water. On one side we pour in pure water, and on the other side we pour in water highly concentrated in salt. Immediately upon pouring in the different water an interesting thing happens. The pure water will pass through the membrane, entering the water with high salt concentration and the water on that side of the membrane will rise. The pressure at which the water level ceases to raise is called the osmotic pressure. At that point, the solution is said to be at equilibrium, or a resting state. The contaminants stay on the same side of the membrane that they were poured into because the pore size in the membrane is too small to let them pass.

Continuing, while this osmosis phenomenon is a fascinating property of nature it really is of no value in the quest to purify water. Indeed, in this first example osmosis sends pure water into the contaminated water, which is not what we want. For drinking water purposes, if we used osmosis, all we would do is simply dilute the contaminants but still drink them. Yuck, who wants that? In fact, what we want is the opposite to happen: we want to extract pure water out of water that has salts and other contaminants in it and send the contaminants down the drain.

With that in mind, our necessity for clean water has facilitated the invention of reverse osmosis. In figure 3 we see that the RO process is achieved by applying pressure to the concentrated side of the chamber and forcing the water backward so to speak, against the natural flow of the osmotic pressure, into the pure water side. What happens now is that we force water out of the concentrated solution into the pure water side of the membrane. Then we collect this water to drink. This action of pushing water flow backward against the natural flow of osmosis is hence called reverse osmosis.

Of course, now that we have created a process, we must build an actual machine to carry out a usable RO process. There are some missing ingredients we need to install into the system shown in figure B to create an actual ongoing process. As our machine is designed now, we would have to manually keep putting in more salt-concentrated water and then scooping out the pure water to drink. Eventually the concentrated side would become too concentrated and the membrane would be plugged up. At that point we would have to manually clean the tank and membrane. I am far too lazy for all that, so let's look to figure C to see how a basic, continually operating RO system really works.

For you and me to design a properly operating RO system we must understand the chemistry, physics and volumes of three different components:

Feed water – The water going into the machine

Product water – The "purified" or processed water

Reject water – The concentrated waste, or reject water that goes to the drain

Accordingly, as you look at figure 3, you will see that we have made three additions to the previous example. We have added the following:

- A feed water inlet line – In a typical home or industrial machine we would constantly feed fresh water, normally tap water, into the RO system so that it could run continuously or fill a storage tank.

- A product line – This is where the purified water, the water we will be drinking, the water processed by the RO, is sent to be collected.

- A reject line with a valve – This is where the wastewater goes and as you will learn, is the most important part of an RO machine. This line is also commonly referred to as a concentrate line and is connected to the drain.

Next, let us talk about how this technology works so that you know both the strengths and weaknesses of it, and more importantly how it might fit into your own water needs. The first thing to consider is the efficiency of an RO membrane, in other words, how much clean water it produces per gallon of contaminated water fed into it.

A critical but little-known fact about RO membranes is that they are for the most part roughly only 17% efficient. What this means is that for a single-membrane system, like most under-the-sink or table-top systems, the membrane will produce only 17 gallons of clean water for every 100 gallons you feed it. Many consumers are frustrated by how much water their home system wastes. Let us see why this is.

One of the things we must remember about RO operation is that as the pure water moves forward through the membrane, the contaminants are left on the feed side of the membrane. If the contaminants can build up to a high concentration, they will form a precipitate, also called scale. This scale will clog up the feed side of the RO system and cause it to stop producing clean water.

For an RO system to operate properly the relationship of feed/product/reject must be precisely maintained. By allowing 87% of the feedwater to go to drain there is proper turbulence, physics, and chemistry to keep the RO membrane clean and happy.

In our figure 4 drawing you will see the addition of a double triangle device placed on the concentrate line. This is called the concentrate valve and it determines just how much water we allow to go to drain. If it is closed, then the rejected contaminants will build up and plug the membrane. If it is wide open, the RO will make extraordinarily little water almost all the feed water will go to drain.

The question then becomes, "If RO wastes 83 gallons for every 100 gallons put into it doesn't that make it a wasteful technology?" Well, yes, and no. Through proper engineering and design, an RO can be made to be 75% efficient or even more. However, doing this takes some care. Here is how we make RO systems more water efficient and green.

The first technique to make an RO system more efficient is called staging, and it is achieved by using multiple membranes. The technique takes the reject (waste) water from the first membrane and feeds into a second membrane. From there it is possible to feed the reject water from the second membrane to a third membrane. By capturing the reject water and feeding it to subsequent membranes, we can reduce how much water the RO wastes (see table 12 below).

In addition to this, it is also possible to recirculate a small portion of the total reject water going to the drain, blend it back into the original feed water and increase efficiency even more. This is called final reject recirculation.

Remember though that staging and recirculating water through membranes has finite limits. At some point the water will become too concentrated for the RO system to handle without the scaling problem we spoke about earlier. An RO system must always have some wastewater running to the drain to transport the rejected contaminants.

KEY WATER DROP – RO membranes merely reject/screen out/concentrate impurities. RO systems do not magically vaporize or destroy contaminants. The contaminants must go somewhere, and that somewhere is down your drain,

After figures 2, 3, and 4 I have provided a small table to show how adding stages and recirculation increases the efficiency of RO systems. Let us pick up our conversation after you have spent some time with the RO drawings.

FIGURE 2

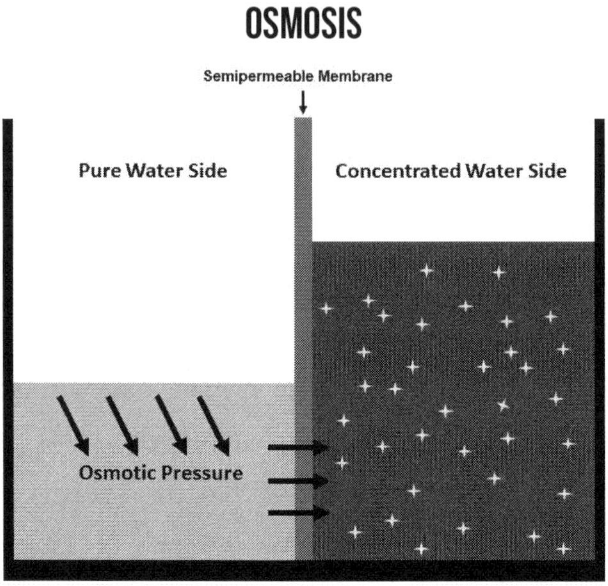

FIGURE 3

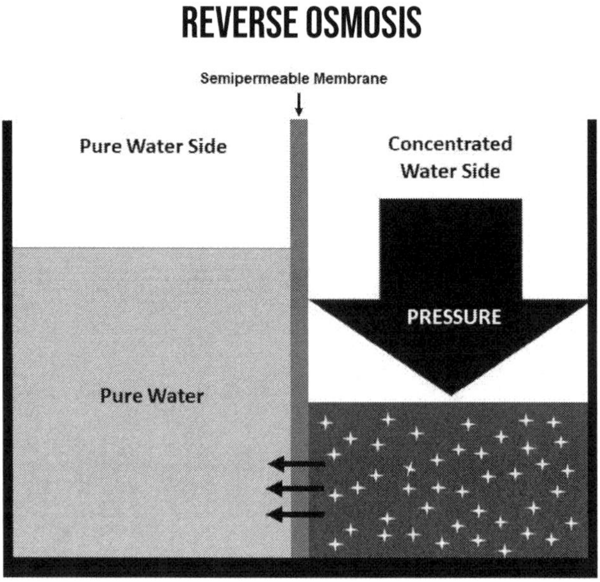

FIGURE 4

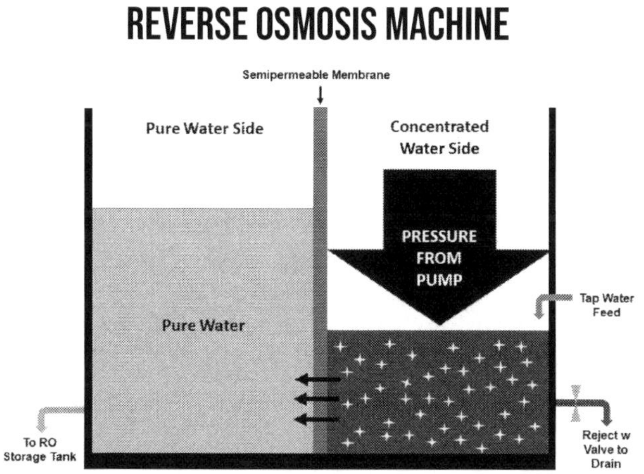

TABLE 13

EFFECT OF MEMBRANE STAGING ON WATER EFFICIENCY		
NUMBER OF STAGES	EFFICIENCY	PURE WATER/ 1,000 GAL
1	17%	170 GALLONS
2	34%	340 GALLONS
3	51%	510 GALLONS
4	68%	680 GALLONS
WITH RECIRCULATION	80%	800 GALLONS

Now, let's look at some additional factors for us to consider when pondering an RO system.

Hardness – In general hardness is considered to consist of calcium and magnesium. While RO will absolutely remove these ions (electrolytes), doing so generally presents a problem. When concentrated, calcium likes to scale, which as we learned earlier will foul/plug the membrane, gradually reducing the rate/flow of production.

This is an excellent place for me to once again jump up on my soapbox and remind the reader just how dynamic, ever-changing, and complicated water is.

When designing a large-scale RO system the engineer must account for the scaling potential of the water. In chapter 4, we learned that in addition to the many tests that were done to provide a water analysis, there were even more tests that the water doctor might have to perform in addition to all of those. We learned about the Langelier Saturation Index Test (LSI), and here it pops up again. The LSI enables us to determine if water will scale the membrane or not. It takes into account, pH, hardness, and alkalinity to determine this. Water may be perfectly safe and delicious to drink, but have an LSI that would scale the RO membrane.

Typically a water softener will be placed before an RO system to remove hardness. In a consumer-level under-the-sink system the vendor usually omits this and just has the customer buy a new RO membrane every so often, which is relatively inexpensive. With a whole-house RO, however, the membranes are more expensive so a water softener is usually prescribed to protect them. If you have a whole-house RO and it is not preceded by a water softener, it should be.

In industry, the other options for preventing RO scale are either the use of an acid feed system, or the infusion of a chemical known as an anti-scalant. Where hundreds of thousands of gallons of water are processed per day the projected costs of acid versus anti-scalant versus water softening can literally economically save or ruin a manufacturing process. Also, in these large industrial settings the amount of sodium and chlorides sent to drain by a softener can present a costly environmental treatment challenge.

On top of this, figuring out the proper amount of harness reduction technology is not cut and dry. Consider this, when RO membranes are staged, the scaling potential (LSI) of the reject water of stage 1 is different from that of stage 2, 3, and/or 4.

The takeaway here is that hardness will scale an RO membrane. In small systems it may be more economical to simply replace the membrane every so often. In larger systems, however, hardness can be a nightmare. So you see once again, even a simple question like, "Do I need to address hardness?" gets the answer, "It depends." The feed water analysis, amount of stages, temperature, pH and gallons used all come into play.

Chlorine – The bulk of RO membranes today are composed of a polyamide/polysulfone material that cannot withstand much exposure to chlorine. Commonly known as thin-film composite and referred to as TFC, these membranes must have the chlorine removed before the water can be fed to the RO. Typically a granular activated carbon filter (GAC) is placed before the RO system.

Another type of membrane known as cellulose acetate (CA) can tolerate chlorine. The problem with these membranes is that chlorine has to be removed after the RO system. Since GAC tends to foster the growth of bacteria, CA membranes are generally not the first choice for drinking water and biomedical/pharmaceutical applications. CA membranes are very effective, however, in certain industrial applications.

Well water, since it is not naturally chlorinated, in general does not require GAC pretreatment.

Sediment – Dirt, sand, colloids and small particles will cause an RO system to plug up. Once plugged the membrane may respond to an alkaline detergent cleaning but sometimes may need to be replaced. For this reason it is generally required to place some type of sediment filtration before an RO system. This filtration can be as simple as a 1-micron disposable cartridge filter, similar to what is available at local hardware stores, or as complicated as backwashing sand filters, chemical treatment, centrifugation, and even membrane filters similar to RO but that are designed for the removal of particles.

In the typical residential application a filter before the RO should still be used. It probably would only need replacing every 6 months and with a cost of only a few dollars it is an easy decision to make. In large municipal and industrial systems, proper sediment removal is a must as these systems have membranes costing as much as tens of thousands of dollars.

Even though a water sample may look clear and be filtered through a 1- or 5-micron size filter, it still may be confounded with great plugging potential by way of particles unseen by the naked eye. Colloidal silica is a frequent culprit and it is not easy to detect.

For this reason, a test known as the Silt Density Index Test (SDI) is generally vital to determine if an RO will plug up or not. It can also help determine what type of pretreatment is needed. An SDI value of 5 or less is the general standard for a tap water RO system. Values of 3–5 are more desirable and values of 3 or less are an engineer's dream. Typically in tap water that comes from an aquifer or well, the SDI can be as low as 1 or less. Tap water that originates from surface water (lakes or rivers), generally has high SDI values and always requires sediment pretreatment.

Pressure – In figure A we see that in an RO situation, the natural tendency of water is to move from a less concentrated area into a more concentrated area. However, to purify water we need the water to move in the opposite direction.

What we want is for water to move from the concentrated side, pass through the membrane, have the contaminants left behind and send clean water into the area of lower concentration, that is, clean water.

This is where the term reverse osmosis comes from.

The natural pressure created in figure A is called osmotic pressure. In figure B we apply a pump to overcome the osmotic pressure and force water backward, so to speak, sending it against the force of osmosis. The pump, when properly sized, allows the RO to operate at maximum output. The pump must be powerful enough to overcome the natural force of the osmotic pressure.

In-home filtration technology, especially in table-top or under-the-sink units the equipment is often offered with or without a pump and overcomes the osmotic pressure with the pressure in the city water line.

For some consumers the output provided by their tap water line pressure will allow the RO to make enough water for their needs. Others, who need more output, generally opt for a system with the pump option. In industrial settings, the RO system always is pump-driven.

It is important to note that the size of an RO pump is not only influenced by the size of the RO system but also by the type of water being fed to it. As the concentration of salts in the feed water increases, so too does the natural osmotic pressure that must be overcome by the pump. Table 13 provides an illustration.

TABLE 14

REQUIRED RO PUMP PRESSURE	
TAP WATER (40-500 TDS)	**30-250 PSI**
BRACKISH WATER (1K-10K TDS)	**250-600 PSI**
SEAWATER >10,000 TDS	**600-1,200 PSI**

Often when we speak of solving the world's water shortage crisis the question arises of, "Why don't we just desalinate more water?"

As you can see from the chart above, a desalinization RO must be operated at very high pressure. This requires an immense amount of electricity thereby significantly increasing the cost per gallon of the water produced, often making desalinization much more expensive than other forms of obtaining potable water. Also consider that desalinization RO units create a large volume of salt waste that must be disposed of. This presents both added cost and environmental/worker health concerns.

Temperature – Unbeknownst to the casual observer the actual output of RO systems is directly dependent on the temperature of the incoming feedwater. More than one unsavvy buyer has been surprised to find out that once installed in their home or business, their RO was not able to deliver the rated gallons per day/hour/minute.

One of the main reasons for this is that the advertised output rate of an RO system more often than not is based on a warm water feed of 77°F. In operation, an RO loses a certain amount of production for every one degree Fahrenheit below that specification of 77°F.

The impact colder water has on RO production is an important consideration when purchasing such equipment. As an example, let us imagine a hypothetical industrial customer that needs one hundred gpm of RO water.

They look at a unit that is rated at 100 gallons per minute (gpm) but soon learn that if that unit is operated at the 55°F their incoming cold water is, that unit will lose roughly 38% of its maximum rated capacity and only deliver about sixty-two gpm. They cannot accept this lower flow rate and have to consider options.

By the way, this output expectation of 62 gpm is arrived at by using scientifically calculated multipliers that are known as temperature correction values.[17] This is how they work:

To achieve the actual output of an RO at any given temperature, we simply divide the rated RO output by the correction value. In this case the correction value is 1.61 and our calculation would be:

100 gpm ÷1.61 = 62.2 gpm

However, since this customer absolutely needs a 100 gpm system, if they have no choice but to operate at 55°F, then they would need to purchase a larger RO that could give them what they want.

The calculation to determine how big of a unit they would need to buy is relatively simple. This is simply a reverse of the previous calculation. In this case we just multiply the desired flow rate by the same correction factor.

100 gpm x 1.61 = 161 gpm

Our hypothetical customer must now make a decision whether to purchase a 100-gpm RO and heat the water to 77°F, or to purchase a larger, more expensive, 161-gpm RO and run on cold water.

Their ultimate decision, however, will not only be based on their cost of electricity to heat water but also their cost of water and sewer service. Uh-oh, you mean there is more to consider?

What we haven't mentioned yet is that at colder feed temperatures an RO is not only less productive but it also sends more water to drain. In comparing the total costs of the two RO systems, our customer must also calculate the higher water and sewer charges that will go along with the bigger RO.

17 https://www.watertreatmentguide.com/temperature_correction.htm.

If the plant has their own well, there is one cost of water. However, if they use municipal water then there is a different, higher cost to consider. If they generate their own electricity their costs will be lower than if they buy electricity off the grid.

Complicated? Welcome again to the wonderful world of water. As I have said before, there are seldom simple answers to even the most seemingly simple water questions. Through examples like this, however, I will help prepare you to get to the bottom of such questions when you are considering your own personal water technology.

The final takeaway in this section is that if you ever consider purchasing an RO for your home, look at the daily production rating and see what temperature it was rated at. Footnote 10 will take you to an actual set of conversion numbers at different temperatures. There are also many similar conversion tables available via an Internet search.

Storage tanks – Almost without exception RO systems utilize storage tanks to capture their water. There are sound engineering and economic reasons for this.

- The longevity and efficiency of an RO membrane are maximized when the discharge of the membrane is allowed to empty to gravity, then shut down and "relax." Emptying into a pressurized line causes the membrane to compact over time. A storage tank eliminates back pressure.

- Most users do not require RO water 24-7 so an RO should be sized for daily consumption and not maximum instantaneous use.

  As an example, a bathtub faucet may fill the tub at 7 gallons per minute, but the entire family consumption is only 800 gallons per day. The cost of a 1,000-gallon-per-day RO might be $2,500 with all the bells and whistles, including a storage tank and pump. Conversely, a 7-gallon-per-minute RO (10,080 gallons per day) would cost about $25,000 properly designed and installed. Storage tanks allow us to size the RO for daily consumption. The tank and pump allow us to meet peak demands.

- Storage tanks allow the RO to start and stop much more infrequently, and operate longer to fill the storage tank as we draw it down. The longer operating period allows the RO to flush itself and make better-quality water. Without a storage tank the pump would be starting and stopping every time we opened a faucet putting great wear and tear on the pump, motor, membrane, to say nothing of severe water hammer on our pipes when the 7-gpm RO kicked on to deliver a few ounces of water for brushing our teeth.

Now that we have presented a veritable engineering manual for understanding how an RO system works, we still haven't answered the questions, "How pure, how safe, how good is RO water?" Let's take a look at the quality of water that RO produces.

KEY WATER DROPLET – RO membranes are not able to guarantee any specific quality of water but instead guarantee and predict rejection percentages. Let's look at what RO removes.

TABLE 15

TYPICAL REVERSE OSMOSIS CONTAMINANT REMOVAL PERFORMANCE	
Aluminum	96-98
Ammonium	80-90
Arsenic	50-90
Bacteria	99+
Borate	30-50

Boron	50–70
Bromide	90–95
Cadmium	93–97
Calcium	93–99
Chloride	92–98
Chromate	85–95
Copper	96–99
Cyanide	90–97
Fluoride	92–95
Hardness	93–97
Iron	96–98
Lead	95–98
Magnesium	93–98
Manganese	96–98
Mercury	94–97
Nickel	96–99
Nitrate	90–95
Orthophosphate	96–98
Phosphate	95–98
Polyphosphate	96–98
Potassium	92–96
Radioactivity	93–97
Silica	90–98
Silicate	92–95
Silver	93–96
Sodium	92–98
Sulfate	96–99
Zinc	96–98

Please understand that this list is not by any means all-encompassing as there are many more potential tap water contaminants and the size of such a list might be more lengthy than this readily available literature on RO performance.

Total lists of RO performance on an expanded list of contaminants are readily available from membrane manufacturers such as Filmtech®. Table 14 is a composite of multiple membrane manufacturers' advertised performance and serves as a very good rule of thumb. For exact removal percentages refer to the manufacturer's specific technical documents.

So now, we still have to answer the question, "How pure, how safe, how good is RO water?"

KEY WATER DROPLET – We can answer this question by using our new water tools. Remember that we learned everything must begin with the water analysis.

So then, to determine if RO is safe, start with your own local or home raw water analysis. Next take a look at the percent rejection numbers in table 14, above. Multiply the concentration of a particular contaminant found in your tap water by the rejection performance of the RO stated in the table, see how that compares to the maximum EPA allowable level, and voila, you can predict how much of that particular contaminant you are likely to find in drinking water if you process is with RO.

Let's work out an example. Let's say that your city tap water is found to have 0.8 mg/l of aluminum in it. This exceeds the EPA recommended limit of 0.2 mg/l and this water would be considered toxic.

However, the rejection table tells us that if we treat this water with an RO, between 96–98% of the aluminum would be rejected. This means if you were to use an RO on this water supply you would reasonably wind up with water having an aluminum concentration of between 0.016 mg/l and 0.032 mg/l, well below the maximum allowable safe drinking water level.[18]

Does this mean that all RO water is safe to drink? Not necessarily so.

Imagine now that your water is contaminated with 12 mg/l of aluminum. If you treat this water with RO you would reasonably expect to have a final aluminum concentration of between 0.24 mg/l and 0.48 mg/l, which would still be considered toxic water.

Would that mean that this RO failed? Not at all, an RO can absolutely be working up to its specifications and not produce water that is safe to drink. If a particular contaminant concentration is too high RO alone may not be enough to produce safe drinking water.

So again, what is the best water technology for you? It all depends on your feed water analysis and the quality levels that you are trying to achieve. We must stop asking what the best filter is and start determining what array of technology is required to deliver safe water.

I am here to tell you, there is no one product or technology that can guarantee safe quality water for anyone. The good news is that you now have the beginnings of a water toolbox to analyze which products and technologies will be best for you.

FINAL RO THOUGHTS

Even with the concerns of pretreatment, pressure, fouling, hardness and chlorine, RO remains the most widely used technology to build a water purification system around. For medical, health care, automotive, marine, food, beverage, military, aviation, general manufacturing and more, RO remains the undisputed champ. Still, it isn't a panacea

---

18 https://www.epa.gov/sdwa/drinking-water-regulations-and-contaminants#Primary.

and it takes a lot of work, engineering and know-how to make an RO system dependable and economical.

I must be honest in saying that I am always cautious of RO systems as consumer products and especially not as whole-house conditioning systems. The caveat for me is that RO systems produce water that is chlorine-free, and they generally fill some kind of pressurized storage tank or perhaps even a water pitcher.

While many like the idea of drinking non-chlorinated water, this is something that should be reconsidered. Some of the worst plagues of humankind have come to us by way of non-chlorinated water. Once chlorine is removed from water, bacteria, algae and fungi are free to grow. A storage tank environment where the water is stagnant and only gets moved occasionally is not much different than a pond. Bad things will grow without having provisions to deal with the standing water.

In a pitcher system, water can be chilled, which slows the growth of organisms. A pitcher can also be emptied, cleaned and disinfected. A storage tank is not so easily sanitized. If you do consider an RO for your home, think about how you will handle the treated water that the RO provides because I promise you, from the time the water leaves the RO machine, it will immediately begin to degrade.

In considering RO for your home, if a tank system is required, seek products that allow the regular sanitization of the tank. If the system simply fills a pitcher or a glass, don't keep the water longer than 24 hours, refrigerate any jugs or pitchers, and wipe the faucet outlet with some bleach mixed 15 parts water to one part bleach.

It is never a bad idea to have a few simple bacteria test kits to monitor your RO water from time to time, and there are some great water test kits available that test quite a few parameters.

Armed with the EPA maximum contaminant chart, the RO rejection table, and a decent test kit, today's homeowner can greatly increase the safety and security of their drinking water.

Enjoy!

7.

Water Purification Technologies – Other Membranes and Filters

Since the advent of reverse osmosis membranes, water technology manufacturers have been inventing and creating all sorts of membranes to filter contaminants. Most of these technologies would never show up in a home/residential application; however, with the world's water becoming more contaminated each day one never knows what the future holds. I thought it would be a good idea to touch on some of this technology so you can keep it in your hip pocket.

KEY WATER DROPLET – At the end of this chapter you will find a graph that describes different contaminants and then organizes them by their molecular weight, called molecular weight units (MWU). When we filter in the membrane range, the MWU scale is a reliable and preferred way of determining which contaminants will be removed by a membrane filter and which ones will not be removed.

In designing water systems, it is not economically prudent to remove particles that otherwise are not harmful to a process but rather only remove the desired particle and any that are larger. An advantage of membranes is that they can be custom designed to remove nothing smaller than the desired contaminant. In this way, the filter will have the highest flow rate and the lowest cleaning frequency possible.

A noteworthy point is that membrane technology is not only used for cleaning water but also a whole host of processes that we wouldn't even normally imagine. A few of them are:

- Concentrating maple syrup and sugars by removing water
- Removing paint solids from automaker paint lines and reusing that water
- Purifying pharmaceuticals
- Recovering undigested food from pig excrement and reducing feed costs (no, I am not kidding)

Nanofiltration – A nanofilter is remarkably like an RO membrane except the pore sizes are designed slightly differently. It is designed to remove calcium and magnesium while leaving many other inorganic contaminants free to pass into the filtered water.

Nanofilters accordingly "purify" water from the standpoint of lowering the TDS; however, they do not deliver the purity levels of RO. Nanofilters also remove organic material as does an RO. In general, nanofilters are often called softening membranes and in the right application can be an economical alternative to salt-regenerant-based water softeners. Nanofilters are too expensive to be used for consumer applications as of this writing. Nanofilters remove particles that are rated in the 300–1,000 molecular weight range.

Ultrafiltration – Ultrafilters have a larger pore size than the previous two membranes and remove both dissolved and undissolved particles. When we leave the range of performance of RO and nanofiltration it is typical that we begin speaking of contaminants in terms of micron size. The reason for this is that as particle size increases it becomes easier and more precise to measure the size of a particle as opposed to molecular weight. An ultrafilter, for reference, is a filter that generally removes particles between the sizes of 0.001 0 0.04 microns. Ultrafilters are commonly used in the medical, biopharma, and electronics industry and do an exceptionally good job of removing biological contaminants such as bacteria, viruses, pyrogens and colloidal silica.

Ultrafilters are more durable than RO and nanofilters and are often constructed of materials like ceramic, which allow them to be cleaned with harsh acids, alkaline cleaners, and super-heated steam.

Microfiltration – These filters employ the screening type of technology that most people envision when they hear the word *filter*. A microfilter is indeed essentially a screen composed of some type of synthetic material that has pore sizes designed to stop particles of a specific size. Microfiltration typically traps particles in the range of 0.05–10 microns. Microfilters are not unlike the common DIY consumer filter that we buy except they are made of higher quality, more durable and precise materials. Microfilters are generally offered with a designation called an "absolute" rating. This means for instance that a micro-filter that is rated as a 0.2-micron absolute filter offers a guarantee that nothing larger than 0.2 microns will pass through it. The common DIY filter is described as a nominal filter and the micron rating is more of an average rating of everything that might pass through.

Microfilters are often used in medical, food, beverage, electronics, power, and clean manufacturing processes. These filters are also often called polishing filters and are used as a final failsafe step to filter water that has already been purified to a high level by other filtration/membrane technologies.

These filters are engineered to a high standard with high-quality materials like medical-grade polypropylene, ceramic, and stainless steel, all of which resist decomposition and sloughing off of fibers into the process water. With ever-increasing water contamination and need for home water filtration, I believe microfilters will soon be commonplace in home filtration products.

Macrofiltration – These filters are used both in industry and consumer filters, performing in the filtration range of 1–100 microns and most often are applied as sediment or particulate filters. As opposed to absolute filters, which guarantee a micron level, microfilters are generally given the nominal rating previously described. Nominal filters are always less expensive than filters with an absolute rating. Nominal filters are generally used early on in a treatment scheme where large, rough particulate removal and not tight specifications are required.

Step filtration – An effective technique in building an efficient water system is to build filtration steps that gradually decrease in pore size as we bring the water to its desired final quality specification. The benefit here is that the cost of a filter increases as the size of the contaminant it can remove decreases. We do not want to ever use a filter pore size that is significantly smaller than the contaminant we want to remove.

As an example, imagine that we wanted to filter bacteria out of a feed water source. We know that a 0.2-micron filter will effectively remove bacteria, so we should just choose the 0.2-micron filter, correct? Ah, not so fast!

Imagine that this water has additional particles that are larger than bacteria. They must also be filtered out; however, these contaminants typically are 1–200 microns in size. If we choose a 0.2 micron only filter, these particles will quickly plug/exhaust the small pores of the 0.2-micron and require frequent filter changes. The proper design then is to place a nominal 1-micron filter in front of the much more expensive 0.2-micron filter. Let the nominal filter pull out the debris and leave the 0.2-micron filter to remove only the smaller particles.

For the sake of argument, a standard 10-inch 1-micron filter can cost as little as one dollar. Conversely, a standard 10-inch 0.2-micron filter can cost as much as $175, far too expensive to be filtering dirt. The lesson learned is, don't ever use more filter than you need, and always use a step-down filtration scheme.

The table that follows provides a visual idea of the size of various contaminants across the filtration spectrum and which filters remove them.

TABLE 16

FILTRATION SPECTRUM			
FILTER TYPE	MICRON RANGE	MATERIALS FILTERED	
MACRO-FILTRATION	5-1,000 µ	COMMON SAND	100–1,000 µ
		LIGHT RAIN DROP	200–700 µ
		SMALLEST VISIBLE DIRT	50 µ
		RED BLOOD CELL	8-12 µ
MICRO-FILTRATION	0.02–5 µ	ASBESTOS DUST	0.2–4 µ
		BACTERIA	0.1–20 µ
		CIGAR SMOKE	0.01-1 µ
		PNEUMONIA BACTERIA	0.04–0.8 µ
ULTRA-FILTRATION	0.001–0.04 µ	POLIO VIRUS	0.03 µ
		FLU VIRUSES	0.002–4 µ
		GELATIN	0.001–0.1 µ
		PYROGENS	0.001–0.1 µ
NANO-FILTRATION	100–2,000 MWU	MILK PROTEINS	
		SUGARS	
		DISSOLVED SALT	
		CALCIUM	
REVERSE OSMOSIS	0–250 MWU	HEAVY METALS	
		ANIONIC SALTS (SULFATE, NITRATE)	
		BACTERIA, VIRUSES	
		ORGANIC CHEMICALS (DYES, DRUGS)	

This table is a conglomeration of many years of my work and several different studies on the particle size of contaminants. For many years the industry standard for matching up particle sizes with filtration technology was a comparison chart provided by a company called Osmonics. I will reference them here and provide a link to the full chart in the footnote that follows this paragraph. I highly recommend keeping a copy or a quick link handy as this is an invaluable water tool.[19]

As we leave this chapter, I will mention that the cutoff micron and molecular weight for various filter designations can vary between manufacturer and scientist's interpretation. In general, though, these ranges hold true across the industry.

In addition, there will always be some variance and range within the particle size of any given substance. Think of it in the same terms as dog sizes for example. How would you describe the size of a generic dog? Chihuahua or Great Pyrenees? Particles and membrane selection are not unlike that. Of course, there is more precise science involved, but no filter, even expensive, highly engineered ones can guarantee 100% removal when the projected particle size is not much larger than the smallest pore size.

19 https://techblog.ctgclean.com/wp-content/uploads/The-Filtration-Spectrumosmonics.pd.

8.
Water Purification Technologies – Filter Media and Backwashing Filters

Perhaps the most universal mechanical water purification technology is something called a backwashing filter (BWF). They are also referred to as pressure filters because they are enclosed and operate on available water line pressure. These filters can be found throughout the world in households, manufacturing plants, hospitals, municipal treatment plants and just about everywhere water treatment is performed.

The operation of a BWF is remarkably simple as the following illustration shows.

FIGURE 6

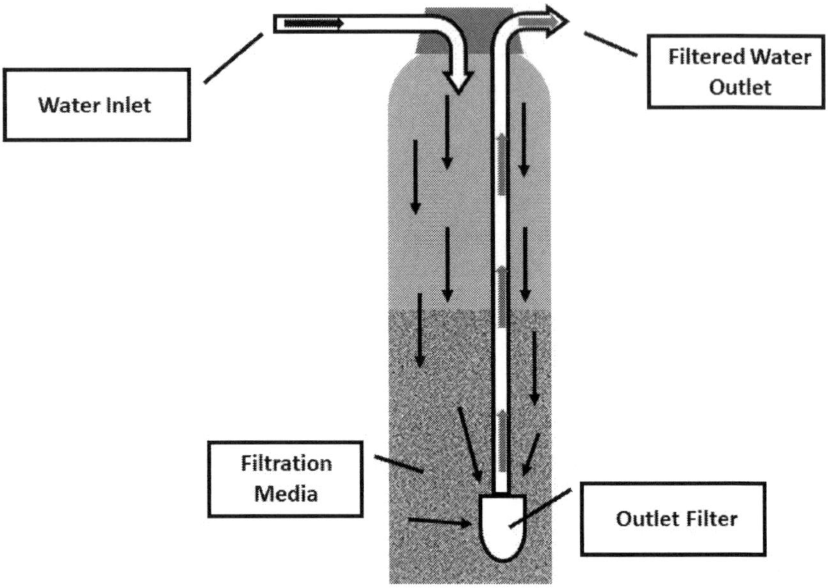

Water enters the filter via a pipe, usually from the available pressurized tap water source (water inlet). Once inside the tank, the water literally flows downward and trickles/passes over and through a bed of filtration media. At the bottom of the filter bed is something called an outlet filter (OF) or distributor. The OF is designed with slots that are engineered to let filtered water pass while at the same time screening the filtration media so that it does not escape and remains in the tank.

Over time the filtration media can become plugged and cause the filter flow to slow or even stop. For this reason the BWF comes with a programmable electronic/hydraulic controller that uses an internal diverter valve to reverse the flow of the water. The outlet line actually becomes the inlet line and the inlet line becomes the outlet. This is called backwashing. During backwash water is fed in reverse direction into the OF. This flow causes the media to expand and fluff up thereby allowing the trapped sediment to be flushed to the drain. After backwash the system returns to normal flow. The following figure represents the backwash cycle.

FIGURE 7

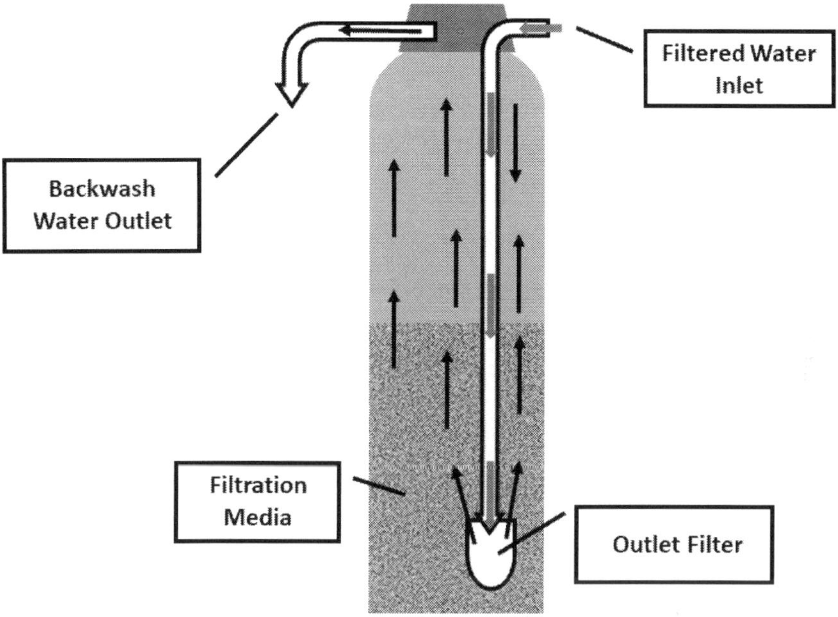

An actual BWF is a bit more complicated than the drawings in that there is a diverter valve that allows fresh tap water to be introduced from the outlet, and also a "T" valve that allows the backwashed water to go to the drain, instead of back into the house water line.

There are quite a few different filtration media that can be placed inside a BF. In general the equipment is the same and it is the media inside the tank that determines what the manufacturer/vendor names the filter. Let's take a look at some of the more popular media that we are likely to encounter.

Filtration sand – As described there are a variety of sand types that can be placed inside a BWF. The main function of filtration sand is to remove dirt, silt, and organic debris from the water. While common playground sand will work, filtration sands are generally finely milled, screened, washed and otherwise conditioned for drinking water and high-purity applications. Filtration sand can generally filter into the 100–200 micron range.

Garnet and anthracite – These are carbon-based "filtration rocks" that can also help filtration sand achieve higher flow rates and better grades of filtration. When used in conjunction with sand, a BWF is commonly referred to as a multimedia filter. Multimedia generally allows higher flow rates, better filtration, and easier backwashing and often less pressure drop than a single media filter.

Green sand and KDF® – These media are called oxidizing media and are most frequently used to oxidize and filter out iron. In both industry and drinking water dissolved iron is a costly problem. In drinking water it not only tastes bad but also as it flows through piping network, the iron comes out of solution (oxidizes) as rust. The rust particles can then clog pipes, valves, faucets, showerheads and more. It will stain sinks and tubs and it can make short work of hot water tanks. In industrial processes it can literally ruin production and inflict millions of dollars in damages.

While expensive technology like RO will remove iron, we wouldn't use an RO for this purpose because iron will foul and plug the membranes.

The good news is that dissolved iron can be easily oxidized before coming into contact with expensive equipment. To clarify, oxidation actually means causing the dissolved iron to bond with oxygen. When that happens the soluble iron forms a precipitate, a sediment, substance that we commonly call rust.

Simply allowing water containing iron to stand, or flow over rocks/grating will actually enhance the formation of oxidized iron. Adding an oxidizing agent such as bleach (chlorine) will speed up the process.

However, for maximum speed in converting soluble iron to rust filtration media such as the ones mentioned above are very effective. This type of immediate iron removal process is desirable in any application where iron-free water must be available on demand, that is to say, upon opening the faucet.

Granular activated carbon – Known to water professionals as GAC, it is undoubtedly the most widely used media in the world. GAC is used in home water filtration products, industrial processes, municipal treatment plants, and even in the cosmetics and personal care products arena. The strengths of GAC are that it removes chlorine, taste and odors, making it ideal for filters that improve drinking water. It can also remove some organic materials like algae and mold, and chemicals such as pesticides and solvents. Its high hardness makes it ideal in back-washable filters where it can be rejuvenated and cycled many times, often lasting 3 years before requiring replacement. The media is commonly used as pretreatment to RO systems where it protects expensive membranes from chlorine destruction. In a later chapter when we examine consumer water filtration products will look deeper into GAC.

GAC works on the principle of **adsorption**, which is not to be confused with **absorption.** The carbon particle does not work like a sponge (absorption), but instead is comprised of countless pores, craters, nooks and crannies. Contaminants actually adhere, or stick to these surfaces by way of something science geeks know to be van der Waals forces. In the backwashing process the debris, and some of the carbon surface, is washed away allowing the GAC to refresh and continue removing contaminants from the next rush of dirty water.

KEY WATER DROPLET – If you have ever wondered where GAC comes from here is a bit of trivia for you. The majority of GAC comes from pulverized coconut shells, which become "activated" by a pyrolysis process that cooks the fragments at 1,000°C or more

(1,832°F). Other sources of GAC are clam shells, walnut shells, corn, rice husks, and virtually all sources of biomass (plant by-products).

Zeolites – Like GAC, zeolites are full of micropores that adsorb impurities. Natural mineral zeolites are an aluminum/silicate-based composition. There are also chemically engineered artificial zeolites. In the golden age of water treatment, zeolites were often used to soften water. Modern technology, however, has rendered zeolites more applicable to special industrial and municipal filtration applications. Their ability to be coated with various chemicals enables them to be used as catalysts in chemical manufacturing processes. Their hardness allows them to be reused. While you and I are not likely to come across zeolites in our home water treatment systems, if we did, they would most likely be delivered via a back-washable filter.

9.

Water Purification Technologies – Ion Exchange Resin

Ion exchange resin (IXR) is also one of the most widely used technologies in the art of water purification. It is the technology found in water softeners and is widely used in many consumer drinking water filters. It is also a highly effective tool for industry, municipalities, and especially the health-care/pharmaceutical/biotech industry.

KEY WATER DROPLET – Earlier in the book we spoke about pure water, noting that it does not and cannot exist in nature. We also discussed that only by water treatment technology could water be made pure. Water cannot be made pure without the use of IXR technology.

While there are some different variations of IXR the most common is something called strong acid cation resin (SAC). This is the resin found in water softeners and is a slight variation of what you will find in many consumer drinking water filters.

SAC is a plastic, gel bead made of a divinyl-benzene-type material. In the processing the bead is impregnated throughout with charged chemical exchange sites. The sites can have either a positive charge or a negative charge, never on the same bead.

- A cation bead has a negatively charged exchange site. This enables it to attract cations, or in simpler terms electrolytes/ions that have a positive charge. These are ions such as calcium (Ca++), chromium (Cr+++), and zinc (Zn++).
- An anion bead has positively charged exchange sites. This enables it to attract anions, or in simple terms electrolytes/ions that have a negative charge. These are ions such as chloride (Cl-), fluoride (Fl-) and sulfide (S2-).

In the example below, the small dots on the bead represent the exchange sites, the attachment points that allow charged particles to be attracted and then chemically exchanged. The idea here is to load the bead with desirable ions and then by passing contaminated tap water over the beads, the undesirable contaminants become held on the bead, while a more desirable ion is released into the stream or drinking water.

Ion exchange works largely on the principle of electronegativity. Simply stated what this means is that the charged bead will favor and trap a more strongly charged ion, and then in its place give off a weakly charged but desirable ion.

Let's look at the figures below and examine how this process actually works in the common water softening application. Our feed water for this example is garden variety hard tap water.

FIGURE 8 represents a freshly charged SAC resin bead. It was washed/soaked in a highly saline solution, in this case clean salt pellets dissolved in water. During this soak, the high concentration of salt causes sodium (Na+) to occupy all of the exchange sites on the resin bead. In a water softener, this is a repeated process called regeneration.

FIGURE 9 represents an exhausted SAC resin bead. In this example enough hard tap water has been passed over it so that it has no more capacity to hold calcium. At this point it is said to be exhausted and if not regenerated or rejuvenated will remain inactive and allow hard water to pass unchanged to the tap.

Notice at the bottom of the figure that the sodium released by the bead passes into the drinking water.

KEY WATER DROPLET – Water devoid of calcium is said to be soft water. The sodium found in the water, by itself does not define soft water, but is the byproduct of using SAC resin to "soften" water. To get the calcium out, sodium is commonly used as the exchange agent.

FIGURE 10 represents a regenerated SAC bead. Once exhausted, washing the bead in an extremely high concentration of salt will override the stronger electric charge of calcium and then replenish the bead with sodium. The replaced calcium is passed down the drain while the resin is ready to go back into service and provide more soft water.

FIGURE 8

Resin Bead Charged With Sodium

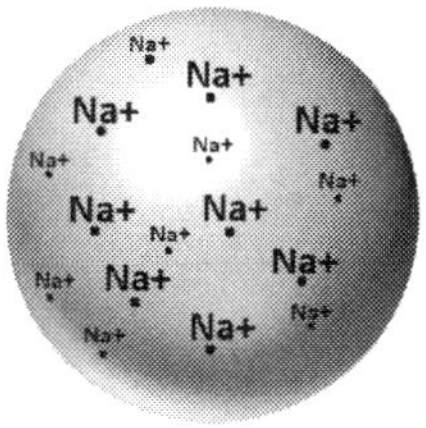

FIGURE 9

Resin Bead Exhausted w Hardness

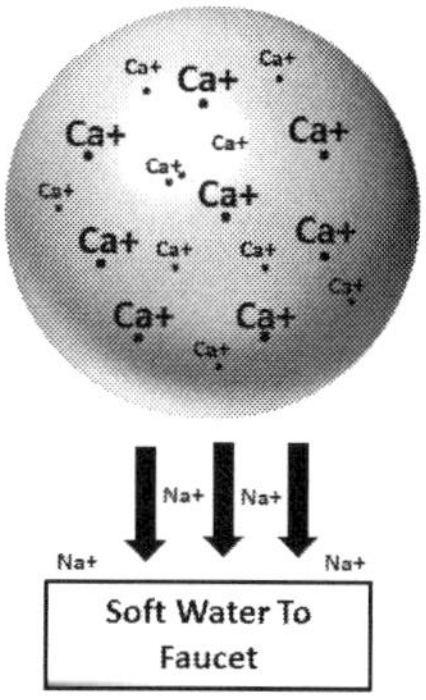

FIGURE 10

Resin Bead Regenerated With Sodium

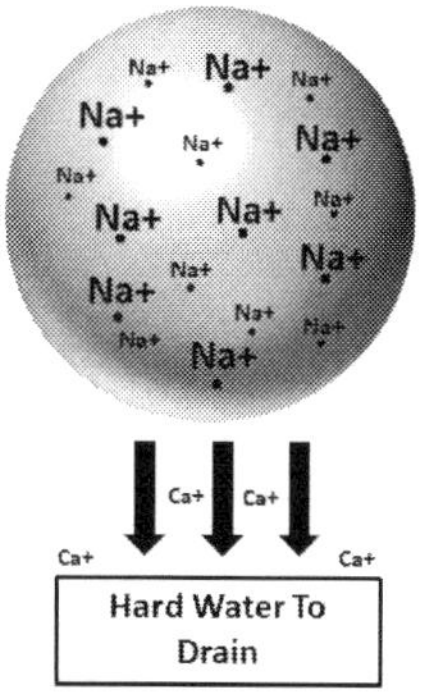

SAC resin exchanges more than just calcium cations. In many tap water applications there may be some iron and magnesium as well as some trace elements. For the most part, however, calcium is the main focus of the water-softening application and the main cation contaminant in all tap water.

In industry there are countless applications for IXR that are not necessarily drinking water-related. It can be used to process drugs and fine chemicals. IXR is also used to remove undesirable ions from electroplating rinse water, automotive paint lines, and even on general wastewater to meet compliance and discharge requirements. We will revisit IXR later when we cover home drinking water projects.

KEY WATER DROPLET – In chapter 3 we mentioned how pure water can only be achieved using technology. In the biopharma/medical/electronics/power generation industries IXR is commonly used to achieve water that is pure from an electrolyte or ionic perspective. It can produce water that is eighteen times purer than triple-distilled water. (Zero Water Filters® achieve zero TDS with IXR.) Let us look at how this happens.

Inside a filter tank we place both cation exchange resin and anion exchange resin. This creates a new product called mixed bed resin (MBR). Figure 11 gives a graphic representation of what this looks like.

- The cation resin is charged with acid, or the hydrogen ion. This is represented as H+.
- The anion resin is charged with the hydroxide ion. This is represented as OH-.

Now let us pass water that is contaminated with zinc chloride (ZnCl2) and see what happens.

- Cation resin captures the positive charge zinc ion and releases hydrogen ions (H+).
- Anion resin captures the – charged chloride ion and releases hydroxyl ions (OH).

Voila! (H+) + (OH-) = H2O. All the electrolytes are removed and all that is left is pure water. Visually the process can be viewed simplistically in the figure below.

FIGURE 11

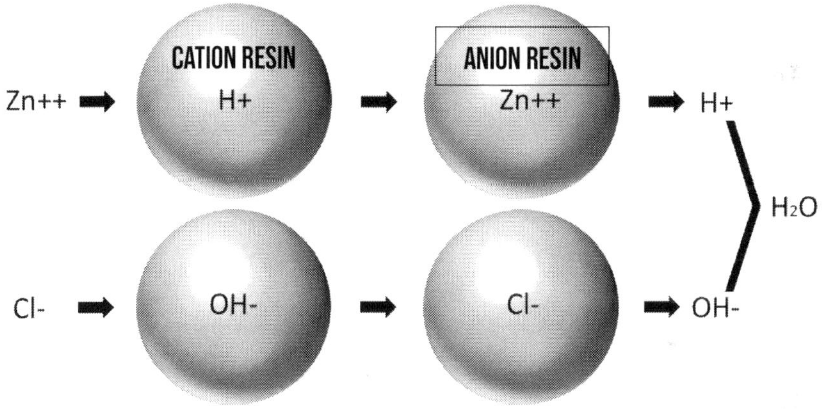

- Typically, to achieve high-purity water, cation resin is charged with hydrochloric or sulfuric acid. (HCl or H2SO4).
- Additionally, to achieve high-purity water, we charge the anion exchange resin with sodium hydroxide, NaOH.

In real-world applications of course, the feed water contains many different cations and many different anions. Mixed bed resin works the same way regardless of the ionic makeup of the tap water.

10.

Water Purification Technologies – Chemical Treatment

The diversity and amount of chemicals that are used in water treatment, if put into books, would fill multiple libraries. In fact, we do have libraries of books just dedicated to chemical water treatment. In the context of providing support tools for your own journey I added this chapter just to highlight some of what goes on, chemically in your drinking water, and in industry.

First off, let us get fluoride treatment out of the way. I am not going to comment much on this contaminant because it is not really a water purification technology but rather a chemical that is added for health reasons. Fluoride is a topic best discussed with your doctor. Later, when we talk about bottled water and home filtration products, we will look at how to remove it.

Chlorine – Chlorine is an effective sanitizing agent that kills bacteria, viruses, and other living organisms in water. To be brief the chlorine molecule captures or steals atoms from certain compounds and replace them with chlorine atoms. Specifically, it is usually hydrogen in the organism's enzyme system that is replaced with chlorine. Once this occurs the organisms simply fall apart and die.

The benefit of using chlorine is that it is indeed effective but also economical, readily available, and easy to disperse in water. One of the downsides of chlorine is that it has a half-life, meaning that it only lasts so long in water and over time will dissipate and no longer keep the water safe. This explains why if you leave an open container of chlorinated tap water outside it may stay clear for a day or two but eventually will be overcome with organisms and turn a nice algae green.

The other disadvantage of chlorine in drinking water is that many consumers are repulsed by the odor and some can even taste the chemical.

One of the unique phenomena of using chlorine in drinking water is that different households on the same city water, even on the same street or water line can experience different levels of chlorine. Many have asked me this exact question: "Why does my water smell more of chlorine than my neighbor a mile away?"

The answer is simple. Because chlorine decays with time and distance, in order to deliver a dose high enough to keep the water safe from organisms to the customer furthest from the water tower, the dose in the tower must be higher than is actually needed.

To illuminate chlorine dosage technique, think about this scenario: it may require the maximum allowable dose of 4 ppm chlorine to be added into the water tower so that a safe level of the disinfectant of 2 ppm remains in the water of the consumer at the furthest line distance from the tower.

However, if you happen to live right by the tower and are one of the first in line to get water out of the tower, then your tap water will have a higher chlorine level than that of the person down at the end of the pipe. The result is that everyone gets bacteria-free water but each user gets a different level of chlorine.

Dosing of chlorine is a constant tiger-by-the-tail for municipalities. On the one hand, water without chlorine will grow organisms and cause consumers to become ill, some fatally so. On the other hand, adding chlorine to water causes a distasteful odor and taste for many. In chapter 11 we will talk about other technologies to sanitize water, but for now, in the US, chlorine is the best choice for delivering sanitary, bacteria-free tap water.

It is important to note that there is a plethora of home water filtration products to remove chlorine, but that technology too has some potential downsides. Later in the book we will also examine the different types of consumer filters.

Chloramine – Chloramine is a water disinfection chemical that is formed by combining chlorine with ammonia. This chemical is effective against many organisms and is widely used by many municipalities to provide safe drinking water. The advantage of chloramine is that it has a longer half-life than chlorine while also presenting less odor and taste. Some municipalities with a low level of organisms in their source drinking water only use chlorine. Others use a blend of both chlorine and chloramine.

The reason a blend of these chemicals is used is that chloramine alone does not kill as many organisms as chlorine does. The ideal blend cuts down on the chlorine smell and maintains sanitary water.

Aluminum polymers – Aluminum sulfate is frequently added by municipalities to prepare drinking water. This chemical causes small particles of impurities to come together and form larger particles. Once enough small particles clump together, the larger particle that forms will fall to the bottom of a tank where it can be filtered or skimmed off. The resultant water then is free of many contaminants. This process is referred to as flocculation or coagulation and is used in waters that have a lot of organic debris and sediment. Typically, surface waters, which include lakes, ponds and rivers, are high in these kinds of contaminants and aluminum addition, or novel coagulants are almost always used. Conversely, groundwater/well water is well known to be devoid of organic and sediment contaminants and aluminum is rarely needed on these waters.

Coagulant chemistry can be used to enhance filtration technology by creating larger, easier to screen particles. Depending on the type of contaminant particle, filtration alone may preempt the need for coagulation. Conversely some contaminant particles will clog, and foul traditional filtration equipment and coagulation may be mandatory. So, welcome to the wonderful world of water filtration problem-solving—a world in which there are no absolutes!

Aluminum sulfate is soluble in water and conventional studies maintain that consumption of the small amounts of aluminum residual that make it into the final drinking water product are safe to drink. Some consumers, however, maintain they can taste something unpleasant in water treated with aluminum.

Leftover aluminum presents a plugging/fouling problem for reverse osmosis systems. It is also very toxic to hemodialysis patients and can cause destruction of blood cells and dementia. Some research is ongoing as well to further determine the possible toxicity of long-term consumption of trace amounts of aluminum.

Acids and bases – The EPA mandates that all drinking water has a noncorrosive nature, with a pH between 6.5 and 8.5.[20]

20 https://www.epa.gov/sdwa/drinking-water-regulations-and-contaminants.

- Scaling water, pH 7.1 or higher, causes calcium and magnesium (harness) to be gradually deposited inside pipes, valves, and fittings over time. A slight scaling potential is desirable.
- Corrosive water, pH 6.9 or lower, causes calcium, magnesium and the metals comprising the inside of pipes, valves, and fittings to dissolve and poison the drinking water. Low pH is what caused the water in Flint, Michigan, to become contaminated. When the pH of the water is corrosive and/or acidic, the treatment plant must add chemical treatment to counteract the threat.

Depending on the raw source of drinking water, the pH of the water before going to the treatment plant can range from 5.5–12, and even outside this range.

If the pH is too high there will be so much precipitation (scaling) that everyone's plumbing, fixtures, hot water tanks, and water lines will eventually clog and block the flow of water. In this case the municipality can add hydrochloric or sulfuric acid to lower the pH to a desirable range.

If the pH is too low, as mentioned, all the metals that water contacts will be dissolved, and the water will become toxic. In this case the municipality will add a base, most frequently sodium hydroxide to raise the pH to a desirable range.

Orthophosphate – For decades now municipalities have injected a chemical called orthophosphate into our drinking waters. The chemical lines pipes and helps prevent lead from leaching into our water. For the longest time this protocol was accepted, and rubber-stamped without much afterthought. However, in recent times, with our heightened concern about water toxicity and environmental impact there are ongoing studies that question the safety of consuming this chemical as well as its negative impact on our ecosystems.

As much as toxicity and environmental impact are important, so too is it important that scientists find a sensible, common ground where we understand that no matter how we treat water, manufacture a product, or develop a new technology, we will have to accept certain downsides. My view is that the furor over orthophosphate is probably overblown, and that the positives of sensible addition to our water far outweigh the potential downsides. We just need to make sure our dosing is accurate and that our record-keeping is impeccable.

Still, there is no definitive position on the safety of orthophosphate from the experts. The FDA believes it to be safe in our foods, while the EPA does not guarantee it being safe in our drinking water.[21]

Biocides – Both industry and municipalities add chemicals to their discharge water to minimize the growth of microorganisms. Biocides are also used to keep cooling water in factories, hotels, and large commercial buildings free of these contaminants.

An outbreak of Legionnaires' disease among convention attendees in Philadelphia back in 1976 revealed profuse bacterial growth in the cooling water stored at the tops of buildings.[22] From that point on, biocides in any type of process water became the rule and not the exception.

Biocides include oxidizing chemicals such as chlorine, chlorine dioxide, hydrogen peroxide, bromine, and ozone. There are also more complex nonoxidizing biocides such as bronopol, THPS, DBNPA and gluteraldehyde. Oxidizing biocides are generally less expensive and work in lower doses. Some conditions, however, require special

21 https://archive.theincline.com/2018/09/27/10-things-you-need-to-know-about-orthophosphate-coming-soon-to-pittsburgh-drinking-water/#:~:text=In%20high%20concentrations%2C%20orthophosphate%2C%20an,fish%20populations%2C%20the%20study%20found.

22 https://www.cdc.gov/legionella/about/history.html#:~:text=Legionella%20was%20discovered%20after%20an,became%20known%20as%20Legionnaires'%20disease.

formulations, and, in these cases, the nonoxidizing products are preferred.[23]

The good news is that chemistry has a grip on killing harmful organisms in non-drinking water areas. The bad news is that most of these chemicals have toxicity and must be accounted for in the waste treatment process. So once again in this book we see that treating water changes it and often clearing up one problem creates another one.

Other chemical treatment – Industry also uses a whole host of other chemicals in their process and wastewater to counteract things like odor, foaming, corrosion and scaling. Like biocides, while these are not drinking water chemicals, at some point in the treatment process they are going to be discharged into the environment. When that happens, as it does every waking moment, our water supplies face continuing contamination.

23 https://www.bywater.co.uk/biocide-water-treatment#:~:text=A%20biocide%20is%20a%20chemical,flow%20and%20under%20deposit%20corrosion.

11.

Water Purification Technologies – Sanitization

Sanitization refers to the removal or nullification of organisms from water. Specifically, we are speaking of bacteria, cysts, spores, viruses, algae, fungi, mold, yeast and basically any living cells that can live and multiply in water. It is difficult to make water that is free of organisms, but it is even more difficult to keep it that way once we store it and begin to use it.

I can never say enough that contamination is the natural state of water. The cleaner we make it the more forcefully and quickly it will absorb contaminants and become toxic.

There are two industry staple technologies that are used to sanitize drinking water and to also keep high-purity water free of organisms. While there are other mechanical and chemical technologies that will kill organisms, the following technologies are the most common in critical applications such as hemodialysis, beverages, food, pharmaceuticals, electronics, power generation and anywhere that water sterility is mandatory.[24, 25] These technologies are also used in many municipalities, even more so in Europe where chlorine is often abandoned in favor of nonchemical treatment. The two technologies are:

- Ultraviolet light
- Ozonation

Ultraviolet light – Ultraviolet light (UV) is a science in itself and we could spend a lifetime learning about it. NASA and the field of astronomy alone can keep you busy enough. Some UV light makes black light posters glow, and others can gently tan or burn our skin. Most UV light has a shorter wavelength than visible light and we humans cannot see it, although some insects supposedly can.

In the wavelength area of 260 nanometers UV light is effective at killing bacteria. It is also very environmentally friendly. UV light is very energetic and immensely powerful. A short visit to Einstein's desk will explain why this is.

We all know (should know?) that the equation for the speed of light is:

$C = (V) \times (\lambda)$

C = The speed of light. All light, UV, visible, infrared, etc., travels the same speed.

V = The frequency of light. Described as how quickly light pulses through one cycle.

λ = The wavelength. This describes the length of that one cycle.

---

24 http://www.ioa-ea3g.org/ozone-themes/ozone-and-its-application.html.

25 https://www.water-research.net/index.php/water-treatment/water-disinfection/uv-disinfection.

Since UV light has a shorter wavelength, the only way it can achieve the constant speed of light is by having a higher frequency. It is the higher frequency that gives UV light the ability to destroy things.

So, getting back to water, many believe that UV light is like a *Star Trek* phaser that zaps bacteria and makes them vanish. This is not how UV works. When bacteria in water are passed in front of a UV field at the proper flow rate and distance, the light attacks the DNA of the bacteria rendering them unable to multiply and carry out basic life functions. Shortly afterward the organisms die.

What we find in high-purity water systems is that storage tanks holding high-purity water are constantly recirculated repeatedly past a UV light to keep bacteria from repopulating. Remember again, once chlorine has been removed from water any new bacteria that enter that water are free to grow and re-contaminate it. Since it is virtually impossible to create a bacteria-free environment in a manufacturing plant, recirculation and sanitization are mandatory.

KEY WATER DROPLET – Using a UV light to reduce bacteria, if applied improperly can cause an increase in bacterial contamination.

How can this be? Well, since UV does not zap or evaporate the bacteria, after passing by the UV bulb the water now contains a lot of dead bacteria. Once dead, bacteria tend to become fragile and lyse, or split into smaller particles. These particles are comprised of protein, carbon compounds, and other molecules that can serve as food for any new bacteria entering the water network. In hemodialysis these fragments are called pyrogens or endotoxin and can create fever and sickness among patients.

In a later chapter, when we talk about home water filtration products, I issue a dire warning to always remember that any time we modify water we create a new set of circumstances in which new contamination can enter the picture. So too, it is with the case of killing water-bound bacteria with UV light.

The good news is that technology allows us to account for this. In high-purity water applications, a UV system is always followed by a submicron filter, approximately 0.2 microns or smaller to filter out all the dead/injured organisms that the light has created.

In noncritical applications, such as swimming pools, hot tubs and decorative water displays, the UV light does not require post-filtration for health reasons as these dead/injured organisms are not toxic to bathers and swimmers. However, since the UV bulb is glass and it is housed in a glass chamber, even in non-drinking applications it is mandatory to have a filter after the UV light. In the case that the bulb or chamber breaks we don't want glass particles going into our pools, tubs, aquariums, and so on. Generally, for this kind of precaution, a DIY 1- or 5-micron filter will keep you safe.

Ozonation – Ozone is the single most powerful water-sanitizing technology commercially available to the public. Ozone is a compound consisting of three oxygen molecules and is represented chemically as O3, oxygen of course is O2.

Ozone has an unparalleled ability to immediately destroy by oxidation the cell structure of viruses and bacteria, thereby causing them to fall apart and die.[26] Ozone is also very safe to use in drinking water because it has an extremely short half-life, which is to say that once released into water it quickly decomposes back into the stable oxygen molecule.

The short half-life of ozone prevents it from being a viable replacement for chlorine in municipal systems. The problem is that it turns back into oxygen before it can travel appreciably far down the water

---

26 https://www3.epa.gov/npdes/pubs/ozon.pdf.

pipeline. Without chlorine in the water, the short-term effect of ozone would allow bacteria to grow in the water once again.

Like UV, ozone can also leave water full of fragmented organisms, which could serve as food for new bacterial growth. Logically then, as is the case with UV systems, a submicron filter of 0.20 microns or smaller should be used after the sanitization to keep the water clean. Also, like UV, ozone is frequently used in a recirculating loop that keeps purified water in a storage tank bacteria- and virus-free.

Ozone is effective in many applications, including industrial and municipal water, and used in the bulk sanitization of the raw water being treated by a municipality prior to chlorination. This is dependent on the design of the individual treatment plant and the specific contaminants found in that water. As with most treatment technologies there is no set recipe for when and where a technology is used.

Interestingly, many communities in Europe do not use chlorine to disinfect water and instead use UV and ozone stations, with enhanced flushing and recirculation techniques located in closer proximity to homes than one would find in a typical water tower design.

Their design philosophy is that killing organisms closer to the home and then recirculating the water before it is delivered to the home is as safe as chlorinated water with the benefit of being chemical-free. Personally, I think the system in the US is safer.

As a final note, both UV and ozone have many applications other than drinking water and high-purity water sanitization. The technologies are used in wastewater treatment, industrial process enhancement, and air purification. For our purposes, we stuck to clean water.

12.

Water Purification Technologies – Boiling Versus Distillation

There are a whole host of contaminants that can enter our drinking water from time to time and depending on variables such as the source of the water, the conditions of the local treatment plants, the surrounding environment, and industrial development of an area, no two cities have the same risk of contamination or the same potential contaminants.

Often, we hear reports of towns being put on an alert to boil water to make it safe for drinking. Boiling is often confused with distilling, or distillation. While distillation does require boiling, it is not the water in the boiling pot, container or bowl that is desirable but instead, it is the steam that when handled properly is a form of purified water.

There is often confusion between boiling water and distilling water and while both processes can help deliver safe water, they are also different processes with different ramifications for the consumer.

Boiling water is commonly done to kill living contaminants, evaporate others, make other contaminants less pronounced/harmful, or even make some contaminants no longer detectable.

On the other hand, distillation, which also requires the boiling of the contaminated water, also involves an elaborate steam condensation/collection and monitoring process and produces a much higher purity water than simple boiling. Let us compare the two processes.

Distillation – In a distillation apparatus, a liquid is boiled, and then a collection tube, called a condenser, is used to collect the steam. Sometimes the condenser tube has an outer compartment to run cool air or water through. This cooling brings the steam down to a cool "condensing" temperature. Then the condensed water runs down the tube, away from the boiling flask, finally emptying into a clean collecting beaker. The water in this beaker is said to be distilled. Figure 12 shows us a typical laboratory distillation setup.

FIGURE 12

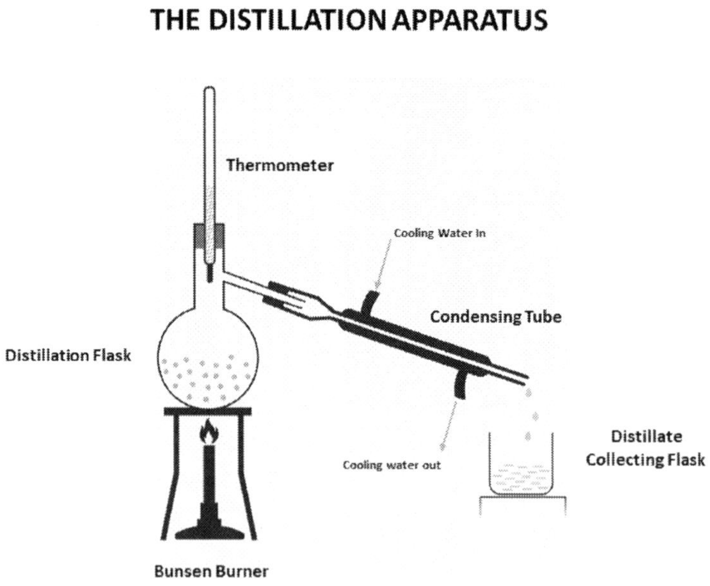

Distillation produces water that is free of bacteria. At the boiling point of water, 212°F, bacteria and viruses are killed, but most importantly are not carried off in the steam.

Interestingly, if there are other liquid contaminants in the flask, and they have a higher boiling point than water, they will remain in the boiling flask and will not enter the distilled water beaker. Conversely, if there are liquid contaminants that have a boiling point below that of water at 212°F, they will enter the collecting flask before the water condensation does, and the distilled water will be contaminated.

This is a critically important aspect of distillation that we must understand. By itself distillation is not a technology that will create pure, clean distilled water. To produce bacteria-free, poison-free water the operator of a distillation device must understand the science of distillation. Distillation works by way of achieving the specific boiling point of a target liquid and then collecting the steam/condensate of that chemical or liquid.

Visualize that we had a mixture of gasoline and water. Distillation would allow the educated operator to separate the two liquids. The uneducated operator, however, would likely collect both refined water and gasoline. Let us see how this works.

Gasoline has a boiling point of 185°F. When we apply heat to the flask the temperature will rise until it reaches 185°F, at that point the gasoline will vaporize, enter the condensing tube and wind up as cooled/condensed gasoline in our collection beaker. Surprisingly, at that point, no matter how high we turn the flame and heat, the temperature inside the flask will not budge from 185°F as long as there is still gasoline in the boiling flask. However, once all the gasoline has been distilled away, the temperature of the flask will rise and continue rising until it reaches the next higher boiling point, in this case 212°F, which is the boiling point of water. Once the temperature of 212°F was reached the water would boil, the steam would rise, enter the condensing tube, condense, and then be collected in the same beaker as we collected the distilled gasoline.

This is important to know for the consumer who would attempt distillation to make safe drinking water. If the composition of a liquid is not known, we could easily distill a toxic, even deadly mixture of what we believed was going to be safe, distilled water.

When distilling any liquid, especially one suspected of contamination, it is critically important to have an accurate thermometer immersed in the vapor stream, as shown in our drawing. If we want to distill and collect only water, then we must measure the temperature in which liquid is converted to gas or vapor. Any liquid not collected at 212°F will be an undesirable and potentially fatal contaminant.

KEY WATER DROPLET – The process of distillation is not a guarantee that the water distilled is safe to drink. The proper process of distillation guarantees that. Companies that sell bottled distilled water are very astute, precise, and safe in their use of distillation. Please, be likewise diligent if you distill water.

Distillation has interesting implications in our lives away from water applications and I think it would be helpful to take a broader look at this technology.

In the petrochemical and refining industry distillation is a staple. There are as many as eleven useful chemical components found in each barrel of crude oil. Each one of those components has an individual boiling point. By knowing those boiling points, the industrial distillation apparatus can separate all these components into separate streams.

Once separated by distillation, those eleven components are next sent to the refining process to create many of the staple products we all depend on. In the world of drinking spirits, distillation is the method by which the desirable alcohol and flavor contents are separated from the undesirable ingredients and by-products.

For IV solutions and injectable drugs, carefully distilled water is highly desired and used, with additional technologies to keep the water bacteria-free once it has been collected.

There are some important downsides of distillation. First off, it requires great amounts of energy to raise the solution to the proper temperature to produce steam. In general, the cost to produce distilled water is substantially higher than a similarly sized RO system. In addition, there is a high degree of maintenance required to keep the apparatus clean and operating properly. Just as other equipment is subject to calcium scale, so is distillation. Making matters worse, calcium becomes less soluble and produces more scale at higher temperatures. Finally, compared to an RO system a similarly sized still will take up a lot more space.

KEY WATER DROPLET – Distilled water is thought by many to be the cleanest, most pure water on earth. It is indeed exceptionally clean water, it is virtually sterile, and bacteria-free. It is safe for human consumption, mandated for many medical procedures and the standard for the manufacturing of injectable pharmaceuticals. However, with all that fanfare, many industries require water that is eighteen times more pure than triple-distilled water. Now, boil that for a minute.

Boiling water – It is not unusual to learn of a community that is put on a "boil water" alert. The reason for the boil alert can be a simple line break in which soil enters the piping network. It can also be because of cross-contamination with wastewater during high flood water. It can be caused by a chemical spill or even by an error at the treatment plant.

Whatever the reason, boil alerts are scary and there is a whole lot of science that we should know when we engage in boiling water. In the same breath I also repeat myself and say that anytime you have a question about your water, the professionals at your local treatment plant are the best source to help you through the crisis.

Know that boiling water is not a panacea. Let us look at some of the important issues associated with it.

One of the things to think about when boiling water is that in the case of bacteria, viruses, fungi, or algae, those contaminants are still going to be in your water after the boiling, and although they will be dead, you will still drink them. In most cases drinking dead organisms will not hurt you, but there are exceptions. Some individuals can indeed be sensitive to drinking certain contaminated water that is boiled.

So, am I suggesting that if your municipality advises you to boil your water to kill certain living organisms you should not drink the water after boiling? Yes, I am saying exactly that.

If there is no other alternative for your drinking water, then boiled water is better than contaminated water. However, if you can afford quality bottled drinking water during a boiling alert, drink that and avoid boiled water.

I reiterate, boiled water is not pure water, it is merely tap water with dead organisms in it. Also, once boiled, most if not all the chlorine will be gone. As we learned earlier, and as I will keep harping on, once chlorine is removed from water, bacteria and other organisms will quickly and freely contaminate it. If you do boil water to drink, refrigerate it, and discard it after 24 hours. Boiled water is free of live bacteria, but it will not likely stay that way longer than 8–24 hours.

Another weak link in the process of boiling tap water is that many harmful chemicals, solvents, and fuels have a higher boiling point than water and will be not be vaporized out of contaminated water that is boiled. They will be left behind in your pot. Conversely, any contaminants with a lower boiling point than water will be removed. The takeaway here once again is that unless we know the boiling points of the specific contaminants in our tap water. Boiling will not necessarily guarantee that the water will be safe to drink.

The following table lists some common chemical contaminants, their boiling points and whether boiling will remove them.

TABLE 17

Contaminant	Boiling Point	Removed By Boiling?
Gasoline	95-395°F	May not all boil away
Diesel Fuel	360-680°F	No
Trihalomethanes (Carcinogenic)	As high as 300°F	Some may remain
Methane	258.7°F	No
Glyphosate (Herbicide)	369°F	No
Common Alcohols	150-171°F	Yes
Benzene	176.2°F	Yes
Carbon Tetrachloride	170.1°F	Yes
Trichloroethylene	198°F	Yes
Toluene	231.1°F	No

As a side note, trihalomethanes (THMs) are frequently cited as a reason for boiling alerts. Over extended periods of consumption, over years, they are suspected to be precursors of cancer. They are enough of a headache today that a short discussion about them is warranted. Let us look at how THMs come to contaminate our water.

Surface water contains high amounts of organic material such as algae, fungi, weeds, grass clippings, trees, leaves and even dead animals. Bacteria naturally decompose this organic matter and, in the process, produce methane gas.

Remember now that municipalities routinely use chlorine to sanitize our drinking water. Chlorine (a halogen) and methane combine to form THMs. The other halogens are iodine, bromine, and fluorine and all combine to form these THMs. Most THMs do boil at a lower point than water, but not all of them do. Boiling is not a 100% safeguard to remove all THMs. As mentioned above, it is a good idea, when asked to boil, to find out which specific THMs are suspected of contaminating your water.

The good news is that boiling alerts generally do not last long, and our potential exposure to THMs is very minimal in these instances.

KEY WATER DROPLET – Believe it or not, for some contaminants, boiling your water can increase their concentrations. In some instances, water that was safe to begin with could be made toxic by boiling. Let us see how this can be.

During boiling, the volume of water in the pot decreases due to evaporation. Inorganic contaminants like lead, chromium, arsenic, and copper do not leave the boiling pot. Now imagine that your raw tap water was contaminated with 1 part-per-billion (ppb) of chromium. If you boiled the water long enough to evaporate half of it the chromium concentration would double to 2 ppb.

An important takeaway when boiling water for drinking is to not boil it for exceedingly long periods so as keep from boiling away a large percentage of it.

The chemistry of water is fascinating, and as I repeat several times in this book, it is as complicated as any chemistry you will ever study. Because we drink it, cook with it, swim, and bathe in it, it is perhaps the most crucial chemistry we will ever study. The idiosyncrasies of boiling are just one example of this magnificent complexity.

With chapter 12 now behind us we will move on to the second half of the book, which focuses on the real-world, everyday things we face with our water. I am excited about moving on to the second half because the topics are more conversational, fun, and common to us. I am also excited because once we get into these topics, and as you journey into your own water situations, the first chapters will also become more interesting, useful and fun.

So here we go, get ready to splash!

13.
Home Water Filters

There are two universally persistent questions that the consumer wants to know about home drinking water.

1) Do we need water filtration in our homes?

2) What is the best home water filtration product?

Let me start this discussion by saying what I have repeatedly said earlier: Our municipal treatment plants and operators do an exceptionally good job of treating our water. Always engage them and become familiar with the dynamics of your personal tap water.

However, the unfortunate reality of modern life is that our water will continue to face ever-increasing threats of contamination. We need not panic, but it is important that we accept home water filtration as an integral part of our safe water future. Here are some important things to keep in mind.

- The world is getting more crowded, we are producing more pollutants and our water supplies are becoming increasingly contaminated. There is no reversing this situation. We have been pumping contaminants into our soil since the Industrial Revolution and we continue to do it today. While municipalities do a good job of removing contaminants, our treatment plants are aging, and pollution concentrations are often so overwhelming that unsafe water at our faucets is becoming more common every day. We must be prepared for this inevitability.

- Of the 75–150 gallons of water per day that the average American uses, only a small percentage of that is consumed in drinking. In the spirit of conservation, we should begin thinking about how our accepting a lower quality water for toilets, showering, laundry, and so on can help reduce the demand on our water resources. We will look at this more closely in chapter 18, A Personal Water Plan, but for now keep the idea close at hand.

- Activated carbon filters remain safe for removing chlorine, pesticides, and many organic chemicals. They are generally inexpensive and easy to install. Many refrigerators are already equipped with these filters.

- The commercial market offers a variety of pitcher-type filters that offer additional purification after the tap such as the removal of metals like lead and aluminum.

- When municipal water does become contaminated, we must demand an advanced, highly visible warning to prevent us from drinking it.

- Bottled water is always safe to drink and a sensible product to have on hand for emergencies.

- Many contaminants that exist today did not exist decades ago when most of our treatment plants were built. Accordingly, the local treatment plant may not be able to safely treat every contaminant.

- Water purification and testing are constantly ongoing processes. A water supply that tests safe at 2:00 p.m. could test to be toxic at 2:01 p.m.

- We must accept the reality that while we do not want to consume any toxic contaminants, whether we like it or not, some of these poisons will make their way to all our food and drinks, but hopefully only in trace amounts.

So now, armed with these caveats let us ask the chapter-opening questions once more.

1) Do we need water filtration in our homes?

2) What is the best home water filtration product?

If it is total water security we seek, then yes, going forward we need some carefully thought out water filtration products in our home.

The answer to the second question, however, is a lot more difficult to answer.

Unfortunately for the average person the selection of filtration systems is never cut and dry but instead depends on many variables starting with things like:

- What is the water source?

- How is this water treated?

- What contaminants are found in the tap water?

- How much line pressure and flow exist in the home?

- What is the daily consumption of the household?

On and on it goes with many other factors influencing the proper equipment for a household water solution.

KEY WATER DROPLET – Always remember, as you will hear repeatedly in this book, any time we filter or alter water, it will exhibit changes in its properties and behavior. Accordingly, our home filtration decisions must predict and account for those changes.

Let us now look at the major types of home water filtration options available to us.

Common faucet refrigerator and pitcher filters – Without having to name product names, I can confidently say that there is an almost uniform similarity in all these products. The main component in these commercially available filters is granular activated carbon filters (GAC), and these, along with filter pitchers are the most widely used home water filtration products. They are readily available, safe, and easy to use. They are a marketeer's dream come true. What could be easier to sell?

As mentioned before, GAC is great at removing chlorine, and certain organic contaminants that can affect taste, odor and even safety. It is important to again revisit the types of contaminants listed in chapter 4 to get a grip on what contaminants we are talking about.

In the world of chemistry, the word organic implies that the contaminant under discussion has carbon in it. Algae, bacteria, methane, diesel fuel, leaves, pesticides, and sugar are all examples of organic material.

KEY WATER DROPLET – Larger in-line cartridge, wall-mounted whole-house GAC filters are generally a technology that is called carbon block. In these filters the carbon used is a fine mesh form that is compressed and bound with an adhesive/binder so that it holds to a formed shape that can then be encapsulated into a usable filter housing. Conversely, pitcher filters are essentially a closed container of granular materials that can freely be shaken up inside the cartridge.

One of the shortcomings of faucet/refrigerator filters is that, as pointed out in previous chapters, GAC removes chlorine. Once chlorine is removed, bacteria, algae and other organisms can quickly begin growing in your water. It is always good to keep filtered water chilled and throw it away after 24 hours. It is also a good idea not to reuse the filtered water container until it is washed in soapy water, rinsed, and dried. Organisms tend to form a contaminant slime on the surfaces of water containers.

In addition, GAC itself is a good breeding ground for bacteria. In its own production water systems Coca-Cola® plants sanitize their carbon filters several times per day with steam made from pure water. Since this is not practical in-home filtration, some manufacturers have upgraded their GAC refrigerator/faucet and even pitcher filters by impregnating them with silver particles. Silver is an effective material for killing bacteria.

Depending on the specific contaminant under consideration GAC may remove it quickly, slowly, or not at all. Herein lies the first shortcoming of current-day faucet/fridge home filtration products. It is also the first obstacle in determining which filter we choose.

Perhaps your water has some methane from nearby fracking operations, while my water may be plagued with odor-causing by-products of decaying vegetation. While a GAC product removes both contaminants, it does so with varying efficiencies. As consumers we really have no practical way to determine how much of a contaminant is being removed, if at all. While it is easy to do a chlorine test at home, testing for complicated contaminants such as methane is expensive and requires the services of an outside lab. Who can afford a $175 per-chemical test every day or every week? The lack of a practical, affordable in-line monitor that can test all possible water contaminants and provide a safe water readout forces both manufacturers and consumers to accept some uncertainty in this process.

KEY WATER DROPLET – It is common for the first liter of water collected from a GAC filter to have some black carbon fines in it. The GAC particle is extremely hard, coarse, and full of sharp protrusions that rub up against each other in manufacturing and in shipping. The fines are safe, and once the GAC is rinsed the fines generally vanish.

The fine print (flimflam?) of home water filters – Challenged by the limitations of which contaminants a filter can remove, many filter manufacturers cleverly provide filter guarantees that simply say things like "reduces bad taste and odor" in water. The small print also frequently may say things like "reduces lead" or "reduces copper," which is not the same as saying that the filter removes the contaminants to safe levels. Here is the caveat: if water is contaminated by six times the safe EPA level of a contaminant and the filter only removes half of it, the resultant drinking water will still be unsafe to drink.

Often, the consumer purchases such products under the assumption that the filter is providing safe, contaminant-free water. However, upon discovering this subtlety in filtration products marketing many consumers feel scammed or betrayed. I completely understand this dynamic and sympathize with consumers who feel they have not been told the truth. This fine print sidestepping upsets me as well.

Before we close on this topic, I would like you to go back and reexamine the Introduction chapter of this book where I stressed the importance of becoming well-educated about water. I cannot say this enough: Our desire to have simple, one-statement answers about what is the best filter and what is the best water is exactly why we remain vulnerable to water products manufacturers that use cleverly worded marketing programs to give us a false sense of security about the filters we use. I believe that it is up to you and me, an educated public, to raise the bar on the language water product manufacturers use to present their products to us.

As we proceed now into other types of filters and aspects of treatment, we will discover that for every seemingly great feature a filtration product offers, there are generally one or more hidden caveats that nobody wants to illuminate. I guess that makes me "nobody" because I do want to illuminate these caveats to you.

NSF, the gold standard or more flimflam? – If you have ever purchased a home water filtration product more than likely it bore the NSF emblem. This NSF, not to be confused with the National Science Foundation, once stood for the National Sanitation Foundation, and today is considered the gold standard by which all filters are measured.

Be aware that this NSF is a private, global corporation that has taken on the personal responsibility for the safety and efficiency of consumer products. Take note also that there is no government regulation of water filtration products, and there is absolutely no requirement that a filter carry NSF certification.

Eye-opening is that an NSF certification is in no way a guarantee that the filter you are buying will deliver safe water. The only guarantee that NSF provides is that the products bearing the NSF logo have passed a series of tests, albeit stringent, that the NSF organization has performed. It is an indirect promise that the product itself is safe, and that it will do what the manufacturer claims.

So again, a crafty water products company can obtain the NSF certification, have a filter that does very little for you, but is marketed with fine print to sound like the water product the world has been waiting for.

Finally, and most importantly, the NSF marking on a water filter does not signify that each individual filter has been tested. NSF testing is not done on each filter but instead is done on a representative batch of product along with a physical audit of the manufacturing plant. However, once the NSF certification is granted, the company does not face a true renewal process for up to 5 years.

I imagine that I am going to take a lot of heat from certain colleagues for the things I am about to say next; however, as the title of my book indicates, my goal is to bring safe drinking water to everyone, and not to pimp or promote any part of the water industry, regardless of my personal interests. That said, I am going to ventilate my thoughts on NSF and its associated certifications.

I am familiar with NSF and can tell you emphatically that they are top-notch when it comes to certifying home water filtration products. With that said, I am also not a fan or proponent of NSF. I will explain.

One of the things that I don't like about the entire NSF process is that it is extremely expensive and makes it next to impossible for small, technically competent companies to get their products into the marketplace and onto store shelves. Most distributors will not accept any water filter that is not NSF certified, even though few could likely express specific details of why they demand NSF products. For sure, NSF has done a great job marketing and convincing users that the NSF certification is mandatory for a drinking water product.

I am an advocate for an entirely new method of certifying drinking water filters, which I believe would provide greater assurance. Think about this: When we are curious about the basic, unfiltered water that comes out of our tap water do we call NSF? We don't, instead we look at our local municipalities' online water report. The reason we do this is that municipalities are made to sample the water they produce, have it analyzed by an EPA-certified lab, and then regularly publish their results.

It is my position that makers of drinking water filters should be required to actually send the filtered water produced by each batch of filters, from several different water sources to an EPA-certified lab, have a full drinking water analysis performed, perhaps on several water sources, and then have a published Internet link to that report so the consumer could see what kind of water their filter was likely to produce.

Care would have to be taken so as not to overburden the cost of filter production; however, the advantage here is that instead of a filter manufacturer undergoing a certification every two to five years, perhaps each batch of filters would have a current performance report. Coupled with our being better educated about monitoring our water, this type of change in regulations would bring us a lot closer to dependably safe water.

Supercharged consumer filters – If there was any good news in the precious sections it is that the manufacturers of consumer drinking water filters realize the shortcomings of their products and too, the testing industry. As a result, they are always striving to find new technology that will give them the upper hand in offering the ideal safe water filter.

Oddly, the filter shortcomings contribute to a perfect storm for us in that we continue searching for a better filter, the elusive perfect filter. This strong public demand creates a market for deluxe models with improved features.

Another downside is that it also provides great financial motivation for every Tom, Dick, and Harry to enter the water market, thereby giving consumers an even more complex and convoluted path to safe drinking water.

So then, here are some of the additions filter manufacturers make to offer us improved, supercharged, high-capacity versions of their standard personal drinking water filters.

- Weak acid cation ion exchange resin (WAC) – This is a specialized form of ion exchange resin that focuses on positively charged ions such as lead, arsenic, copper, iron, mercury and more.

In understanding this media, I suggest a quick revisit to chapter 9 where we spoke about ion exchange resin (IXR) and more particularly about strong acid cation resin (SAC).

Many companies add WAC to their granular activated carbon (GAC)-based filters to add the capability of toxic metals removal to their already strong combination of chlorine removal, taste, and odor improvement.

We learned earlier that strong acid cation resin (SAC) was the preferred IXR for removing calcium and magnesium hardness and producing soft water. Note that calcium and magnesium like the other toxic metals just mentioned are also cations. It is important to note that in addition to removing hardness, SAC will also remove the toxic cations under discussion.

So, the question begs to be asked, why does the consumer water filter manufacturer not just use SAC since it removes a wider range of cations than WAC?

The answer lies in chapter 5, where we learn how to read a water analysis.

Across the globe if we look at a typical drinking water analysis, we see that the overwhelming majority, often 99% of the cation contaminants, is calcium and magnesium, otherwise known as hardness. Indeed, the overall hardness of tap water can be 100–500 mg/l while the total concentration of all toxic cations is usually 2 mg/l or less, even far less. The hardness ions are not harmful to health while even in low concentrations the toxic cations can cause severe health problems.

Segue now back to this conversation of WAC versus SAC.

If a consumer drinking water filter were loaded with SAC, 90% or more of the capacity of the filter would be consumed removing harmless cations instead of harmful ions. The result is that the consumer would be changing filters every 1, 2, or 3 gallons. At a cost of \$3–\$5 per filter the cost of our drinking water would escalate to \$2 or more per gallon.

The beauty of WAC is that it does not remove the harmless cations associated with hardness, and instead primarily removes the trace, but toxic, cations. In this manner the filter lasts much longer and can achieve the desired capacity of 40 gallons per filter or more.

If you ever decide to cut open one of the common and standard drinking water/pitcher filters to look inside you will find a combination of black particles (GAC) and amber or greenish particles, which will be the WAC.

As a final note, for those who have hard water and wonder if they can purchase a cartridge filter to remove the hardness, I would answer that while you probably can find such a product, it would be prohibitively expensive to use.

For hard water, a water softener uses 15 pounds of salt per regeneration and delivers an average of 2,500 gallons of soft water. Fifteen pounds of salt costs about \$2 when purchased in 50-pound bags. This translates to about 0.08 cents per gallon. Conversely, the cost per gallon of soft water provided by a disposable hardness removal filter is going to be more than \$1 per gallon.

The takeaway here is that for removal of toxic trace cations in a home water filter, WAC makes good technical and economic sense. On the other hand, if soft water is the goal, a traditional water softener with SAC is preferred.

The shortcomings of WAC that the consumer should know about are first the aesthetic or social concerns. WAC is made of a plastic-type material. As such it requires petroleum as a base ingredient and in the home filter market winds up as more plastic going to landfill.

Also of note is that IX resins never really stop working, they are always constantly exchanging one ion for another. In contaminated water the science is such that WAC will hold on to toxic ions and release a harmless one that was used to charge it during manufacturing. However, once the resin exchange sites are completely occupied with toxic contaminants, it then tends to regenerate or rejuvenate. In other words, it can start holding on to the harmless cations that it once ignored, and shift gears to release the toxic, undesirable cations that it initially removed. For this reason, it is important that we do not operate the filter past the recommended gallons predicted in the instructions.

- Mixed bed ion exchange resin (MBR) – MBR is a combination of cation resin (CR) and something we have not yet talked a lot about, which is anion resin (AR). In general, filters that use this material often boast about producing 0 TDS, or even pure water that is free of contaminants.

In nature, in physics and chemistry we find that there is always an exact balance of cation and anion contaminants in water. A cation must always be associated with a corresponding anion because opposite electrical charges in water are always satisfied, equally balanced with each other.

In water the cations sodium, calcium, copper, zinc, and so on are paired with anions such as chloride, carbonate, and sulfate. Generally harmless, CR will encounter a compound such as lead chloride, capture the lead, and give off something like calcium, which then combines with the available chloride to leave us with harmless calcium chloride.

Nonetheless, even when ions are being exchanged, there must always be an equal amount of cation charges and anion charges in any water specimen.

Note that I did not say an equal amount of cation and anion particles but rather an equal number of charges. Water can never have an overall electrical charge, so all the charges must have an opposing and equal charge. In the case of calcium, which has a positive charge of +2, it would take two particles of chloride, which each have a charge of -1, to achieve electrical charge equilibrium.

So then MBR removes both positively and negatively charged particles. It removes both cations and anions, and the result is water that is theoretically free of ions. For reference as to whether ion-free water is pure water take another look at chapter 3 and our pure water discussion.

The shortcomings of MBR include again, as with all ion exchange resins, that these are made of plastics. In addition, as the anion resin is used and approaches exhaustion it often gives off a fishy odor that many users have discovered and objected to.

Finally, there are some important shortcomings of so-called deionized water, also called 0 TDS water.

- 0 TDS water is very hungry for available ions. Over long periods of time, especially in young children it can pull calcium out of teeth.
- 0 TDS water is unstable and will quickly degrade if left standing. Particularly it will pull CO2 from the surrounding atmosphere, form carbonic acid, and lower the pH of the water to as low as 5.5. While not harmful, the water can taste sour or metallic.

The real caveat with the pure water boasts is that we do not need to drink water that is pure or devoid of all contaminants. Some contaminants are safe and even beneficial. All we need is water devoid of harmful contaminants.

- Titanium dioxide (TiO_2) – This is another adsorbent material that unlike GAC has a strong ability to remove toxic materials such as lead, copper, and mercury, while also providing an ability to reduce bacteria. More than one of the leading consumer water filter manufacturers produces a filter using this material. While generally considered safe, there is ongoing research indicating that small particles of TiO_2 enter the human body, which might cause negative effects to cell structure.

From a commercial standpoint, filters utilizing TiO_2 are also generally more expensive.

Home reverse osmosis systems (RO) – RO technology is considered by most vendors as the Cadillac of home water filtration. Chapter 6 goes into great detail about this technology but for this chapter we will focus only on the implications it has in our homes.

The number of vendors, brands and styles of RO systems are as varied and numerous as the mind can fathom yet there are two basic systems that we can purchase.

- Under-the-counter/countertop RO systems
- Whole-house RO systems

As we learned earlier, a great feature of an RO system is that of all the purification technologies, it offers the widest span of the types of contaminants it will remove. Literally, an RO system can make safe drinking water out of virtually any feed water whether it originates from industrial waste, radioactive water, and even municipal sludge. Let us just not put RO to those tests, OK?

Even with its amazing contaminant removal abilities, RO is not a panacea, and can create just as many problems as it alleviates. Some of the caveats follow:

- For proper membrane life, RO feed water must be softened. This adds both initial expense and operating expense for a water softener, and salt. An RO can be run on hard water, but the membrane will scale and require frequent and costly replacement. Iron will also quickly foul an RO membrane.
- Small RO systems, those making 100 gpd or less, can send to the drain as much as 83 gallons for every 17 gallons of clean water consumed.
- RO water tastes flat to most people.
- RO water is devoid of chlorine and readily grows bacteria as it sits.
- RO is prone to absorbing CO2 from the surrounding environment thereby forming carbonic acid, which can cause the water to taste metallic, and the pH of the water to drop into the 5.5 range.
- The lower in contaminants (TDS) the RO feed water is, the purer and more aggressive the RO product water will be. In areas with exceptionally low TDS feedwater, the water produced by RO, in a whole-house water system can leach lead and copper out of the plumbing system and contaminate otherwise safe drinking water.

Most home RO systems include a storage tank of some type. The reason for this is that it is more economical to design a smaller RO to handle total daily consumption and use a storage tank/reservoir to deliver the short-term maximum flow rate needed to handle the opening of a spigot, a shower or blast from the sink.

Here is the thought process behind RO; forgive me if this is redundant.

An entire family might use 750 gallons of water per day, over a 24-hour period. However, during high use they may require 10 gallons per minute for a short duration, let us say 10 minutes.

The cost of a 10-gallon-per-minute RO system (14,400 gallons per day) could easily cost $40,000, occupy many square feet of space, and send the electrical bill skyrocketing. However, the cost of a 1,000-gallon-per-day RO with a 1,000-gallon storage tank and 10-gpm pump could be had for as little as $4,000 and be far less space- and utility-consuming.

KEY WATER DROPLET – An under-counter RO system is going to reduce your kitchen storage space and require that a hole be drilled through your countertop to accommodate the RO water faucet.

Wall-mounted cartridge systems (WMS) – There are a variety of available water filtration systems that use disposable cartridge systems to deliver varying degrees of drinking water treatment. These can range from a single standard filter housing mounted where your water enters the house, to well-designed, attractive wall-mounted systems intended to be placed before coffee makers, drinking fountains or even sinks. The cartridges offered generally include combinations of activated carbon, sediment filter, scale reduction, odor removal, bacterial filters, and mineral removal.

The advantages of these types of systems are many. Some do not require electricity or a pump, they are lightweight, take up wall space instead of counter or cabinet space, are user-friendly, and can be chosen to cater to specific water contamination problems. Also, by only treating water intended for drinking they are more economical and less maintenance intense than whole-house systems.

The disadvantages of these systems include the caveat I speak about often, which is, they remove chlorine and allow water to stand still. These conditions harbor bacterial growth inside the filter. For this reason, it is important to use these systems regularly and change cartridges often, maybe even more frequently than the manufacturer recommends. I am sure the manufacturers of these systems love me now! I promise you, however, my recommendations are based on your safety, not on helping to sell someone's filters.

Salt-less water conditioners – First and foremost it is important to call these devices water conditioners and not water softeners. A water softener removes calcium and replaces it with sodium to produce what is called soft water. The devices we are talking about here do not remove anything. Instead, they deliver an electrical force that causes the calcium (hardness particles) to become excited, therefore remaining in solution so that they do not form scale inside your hot water tank and plumbing fixtures.

Do these things work? Unfortunately for the consumer, yes, they do. Please allow me to explain.

The science behind using electricity to prevent ions from precipitating is indeed real science. Precipitation by the way, is the proper term that describes any ion (mineral) coming out of solution to form a deposit. Commonly we refer to ion precipitation as scale, or rust.

The problem with this version of science, and really with all of these type of products is that once the ions, in this case calcium, have been excited with electricity, they immediately begin to release that energy and return to their normal state, which means they will go back to precipitating in your pipes and fixtures.

If these systems electrified the entire length of your pipes, then they indeed would have a high probability of stopping calcium scale. This of course would be expensive to operate and to install as all your pipes would have to be exposed to electrify them.

The makers of this technology would maintain that since copper conducts electricity, the entire piping network is indeed activated by the energy applied by the device. Again, this is not altogether false, but it is contradicted by one indefatigable law of nature that casts severe doubt on the potential effectiveness of this technology. Ions, and all things in nature.

The laws of thermodynamics teach us that all matter, all particles, ions, objects, and compounds constantly seek to expel energy and reach a resting state called equilibrium.

To illustrate: If we hold a ball in our outstretched arm, the force of gravity pulls on the ball. The ball "wants" to fall, but the energy of our arm holding the ball overcomes the ability of the ball to fall. However, if we open our hand the ball will fall to the ground and it will experience a series of bounces until it comes to rest, still, motionless on the ground. At this point, the ball has released all the energy it was holding into the surrounding environment as heat and sound and remains at rest.

In the same way that the ball releases the potential energy given to it by elevation, so too will the calcium ions excited by electricity eventually release that energy as they travel down the length of and/or sit idle inside your house pipes. Like the ball, the calcium ions will quickly return to their normal state, which is a scaling state.

So then, the question at the end of the day becomes, "Do these filters work?" The best way I can answer this is to say that based on my grasp of science and water technology, I would never recommend this equipment and would always choose a water softener. The technology is not bogus, it is authentic, and the suppliers provide safe, well-designed equipment. Still, with all of that, I do not believe the non-scaling effect provided lasts long enough to be effective in the home.

Water softeners – By far the most familiar home water treatment technology is a water softener. Frequently installed in basements and garages water softeners remove calcium, manganese and iron from tap water and deliver water that is easier on pipes, fixtures, and hot water tanks.

Softeners work by the process of ion exchange described in chapter 9. In this system cation IX resin is charged with sodium ions, which come from a salt (NaCl) charging. The resin has a stronger affinity for calcium, so it holds it and in its place releases calcium. Hence the name ion exchange. We exchange one ion for another.

Sodium is considered a desirable ion because it is very soluble in water and does not easily form a precipitate/scale. In addition to being friendly to pipes and hot water tanks, soft water greatly decreases the amount of soap needed to wash clothes. Soft water also eliminates the soap scum, commonly called curd, that builds up around a bathtub at the surface of the water. The deposit that hard, soapy water leaves was once commonly called a bathtub ring. Soft water can also virtually eliminate nagging spots on dishes, windows and washed cars.

As a simple rule of thumb, a softener will give off 2 parts of sodium for every 1 part of calcium it attracts and holds. This can have implications for people who maintain low-sodium diets; therefore, it is good for individuals on such diets to check with their physician before installing a whole house water softener. Let us look at how a softener might affect our diet.

- Typical tap water has about 150 mg/liter of calcium.
- If we soften this water it will have about 300 mg/liter of sodium.

For the average American most health sources recommend a maximum daily sodium intake of 2,000 mg per day. So, given the conditions above, a consumer would have to drink 7 liters, nearly 2 gallons of water to exceed the daily maximum target of sodium.

Clearly, in normal cases sodium from soft water is not that much of a threat to low-sodium diets. However, it does have some impact, and if the incoming tap water hardness is higher than 150 mg/l the amount of sodium in the drinking water could be two or even three times higher than in our example. With this information at hand, it is easier to decide whether a softener is a good idea for you.

So, what about the individual who has terribly high hardness but also wants to limit salt intake? Fortunately, there are a couple of ways to get around this challenge.

- **The "bypass and blend" solution** – This is a clever way to use a water softener to lower your total water hardness but not eliminate it. The proper installation splits your feed water line into two separate paths. One path goes directly to your house plumbing, but the other path is sent into a water softener.

  Then, using a blending valve, the two waters are combined to achieve a desired level of reduced hardness. When installed properly the level of hardness is reduced to where scaling is minimal while the amount of salt put into the overall water is not harmful to those on sodium-restricted diets.

  In general, a hardness of 50–90 mg/l (ppm) will prove to be relatively non-scale producing. A simple schematic of this system is displayed in figure 13. Note that the water streams split into two before the blending valve, and then recombine after the check valve. The check valve prevents water from backing up into the softener. In some cases, a pump may be needed to drive the flow forward. Also, a common shower blending valve that regulates water temperature by blending hot and cold water together can be used to achieve this design.

FIGURE 13

SOFT WATER BLENDING APPARATUS

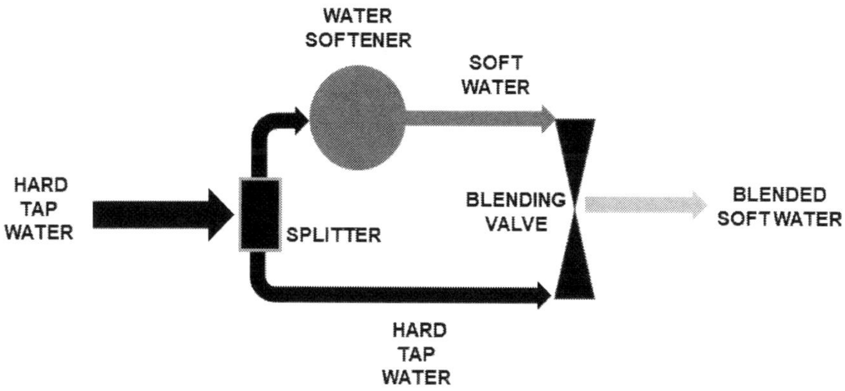

- **Hot water tank only softening** – The costliest household damage that hard water causes is premature hot water tank failure due to scaling. While the appearance of scale buildup in our showerheads and sink fixtures is also something we would like to eliminate, there is another way around the scaling problem that doesn't require us to soften 100% of our tap water.

  By softening only the water going into the hot water tank, the tank will last forever, and as an added cool feature, when you take a shower, you will essentially be getting the blending benefit mentioned in the previous bullet point. The difference of course is that the greater the amount of hot water you blend into your usage, the softer the water will be.

  There are a couple of advantages to this option. First off, hot-only soft water will reduce scale buildup in your fixtures. Secondly, the sodium in your hot water system will act as a solvent to help calcium from building up.

  The other advantage has to do with the characteristic manner that soft water feels on your skin when in the shower or when getting out of the tub. Soft water feels slippery on our skin. Some folks do not like this at all, and for many it takes some getting used to. If you use one of the two techniques described here, your water will not feel slippery.

Again, the determining factor in whether to use a softener or not, and whether to implement a blending system depends on the individual. In high hardness applications a softener may be mandatory. With sodium-restricted diets the blending concept offers some genuinely healthy benefits.

Iron filters – Iron is a nagging problem for many Americans. Largely resigned to well water, iron causes red staining of tubs, sinks and fixtures; it fouls and ruins hot water tanks; stains clothing; leaves spots on dishes, windows, and vehicles; and makes water smell and taste bad. Anyone who has iron in their drinking water does not have to be told or see a test for it. The telltale signs are universally infamous.

Iron exists in two chemical states: The ferrous form is clear and dissolved in water. The other form, ferric iron, is the oxidized form that we commonly refer to as rust.

The overwhelming majority of iron-removal technologies involve treating the ferrous form with an oxidizing agent that will convert (oxidize) the contaminant to ferric rust, and then removing it by screening/filtering it out using gravity and some form of cartridge technology.

Ferrous iron can be oxidized in a variety of ways. Chosen methods are often defined by economics, product availability and existing filtration technology. The following are a few additional thoughts regarding the treatment of iron in your tap water.

- Simply stirring/sloshing/aerating iron-contaminated water will eventually cause the iron to precipitate out, i.e., form rust.
- Adding 2–5 mg/l of chlorine (bleach) to the above method will enhance the oxidation process.
- A back-washable filter containing manganese greensand media, regenerated with potassium permanganate, has been a staple filter for oxidizing iron for many years and is still widely used today. This technology is somewhat messy and can leave pink/purple stains around the water system, as well as a pink tint in the resulting water. It is, however, 90–100% effective at iron reduction.
- A back-washable filter containing one of the newer oxidation media such as KDF® can be charged with chlorine, provide excellent iron reduction and overall operation that is neater than a permanganate system.

KEY WATER DROPLET – Water softeners are effective at removing small amounts of iron but are increasingly subject to iron fouling and contamination with iron-feeding bacteria as the iron concentration increases. As a rule, once the iron content reaches or exceeds 0.3 mg/l or .3 ppm, an alternative technology is preferable.

KEY WATER DROPLET BONUS – Rotten egg smell in water, caused by hydrogen sulfide, can also be reduced and/or eliminated with an oxidation system. Chlorination is especially effective as are some forms of ultraviolet light treatment.

One of the challenges we have in choosing home water filtration is that both opportunistic and ethical water product manufacturers are inundating us with new home filtration products at a pace that is hard to keep up with. I wish I had the time and resources to do a complete review of each of them, but such a task would be next to impossible to perform thoroughly. Nonetheless, the information provided in this chapter, fortified by the previous chapter will enable you to ask effective questions and hopefully choose products that are truly best for you.

Key home filtration takeaways

- Regularly download your local water report and know your contaminant risk.
- Get to know your local water plant. Schedule a visit and check out what they do.
- Use a high-quality faucet filter and/or refrigerator filter.
- If you need greater assurance use a high-quality water pitcher product. Just make sure you do not keep the water longer than 24 hours.
- Stay away from whole-house water filters. Unless you are plagued by iron and high hardness, you do not need a whole-house filter. Toilets, showers, windows, boats, decks, and cars are all fine with your city tap water, no matter where you are.
- Focus on and treat the smaller volume you drink. This will minimize your expenses, time and worries about water.
- Seek the NSF seal but do not be repelled by the lack of it. Make sure the filter you are considering above all, satisfies your new water knowledge base.
- When choosing a vendor to install a system in your home, look to established names with a long history of safe customer service. Companies/brands such as Kinetico®, Culligan®, WATTS®, Fleck®, and Rainsoft® have been in the business for a long time and provide a wealth of technical knowledge.

14.
Bottled Water

During the snowy Christmas season of 1986 my then brother-in-law told me that I should start a company and sell pure, safe, delicious water in bottles. I told him that he was out of his mind, elaborating that nobody would ever pay a premium for something that was virtually free out of the faucet.

Regardless of which market report we look at, it is generally agreed upon that the annual global market for bottled water is fast approaching $300 billion. Furthermore, the global concern for safe drinking water is expected to keep pushing sales of these products upward with no leveling off or downturn on the horizon.

Had I listened to my then brother-in-law, I would likely have more money than the Federal Reserve Bank today. I think I should revisit him and get some sports predictions. Wow, did I miss the boat or what?

Please excuse the pun, or play on words, but our thirst for safe, clean, tasty water has created explosive opportunities for entrepreneurs and beverage companies to entice our dollars with new, exciting, secure water products. We all want to find that perfect water: Ponce de León's Fountain of Youth, if you will, and there is no shortage of products offering us exactly that.

The reasons why people purchase bottled water vary greatly. However, fueled by a lot of bogus/frivolous/manipulative marketing and sales claims a great number of consumers are buying bottled water to receive health benefits that bottled water products simply cannot and do not provide. The good news in all of this is that for the most part commercially bottled water is safe.

Let us review the more popular reasons why we purchase bottled water.

- Convenience when traveling, attending major events, picnics, work, camping, and more
- Backup supply in the event of tap water contamination
- Better taste than our own well water or municipal tap water
- Nutritional benefit
- Benefits to health and well-being
- Distrust of our municipal tap water

As you might expect, bottlers offer a broad spectrum of bottled water products to attract the different types of customer demands that make up this immense market. With that said, let us also review the various types of bottled water that are available for our consideration.

- Premium processed drinking water – These are the most popular and widely purchased water products. These products are nearly identical to each other. They are all municipal tap or well water that has been meticulously processed, reengineered, and offered to the consumer through grocery stores and virtually all convenience and major consumer retail outlets. Products such as Dasani®, Aquafina®, and Nestle Water® are some of the more famous players.

- Specialty springs and aquifers – Second in popularity are what some refer to as the natural waters. These are waters that come to us by way of natural artesian springs or underground wells called aquifers. The bottlers of these products each boast of mineral, taste, and supposed health benefits that make their individual brand of water the best water on earth. Fiji Water®, Starkey Springs® and Poland Springs® are a few examples.

- Sparkling water – Sparkling water is a staple of the customer who flocks to the appearance of class, style, elegance, success, and exclusivity. I imagine that many consumers also simply like the taste of sparking water. Editorial comments aside, sparkling waters are those providing effervescence or bubbles. This is done either through naturally occurring carbon dioxide such as in S. Pellegrino®, or via the injection or reintroduction of carbon dioxide into the water, as in the case of Perrier®.

- Fortified/enhanced water – A relatively new group of products, these companies take a basic water, which more often than not is purified tap water, and then add a unique, patented recipe of chemicals that promises to improve everything from our health, energy, to our general well-being and more. Vitamins, minerals, flavoring, pH adjustment, hydrogen and more comprise the water enhancements companies may choose from in creating their concoctions.

With all the options of bottled water available to us trying to decipher the differences and choose the perfect and ideal bottled water can cause a migraine. The advertising claims and customer testimonials only seem to make the choice more confusing.

The worst part, in my observation, and something that brings me great unrest, is that all too often consumers bring home bottled water products and then still have doubts and questions as to if they have made the proper water choice for the ones they care about. Let me be clear, no consumer should ever have doubts or uneasiness about the water they drink, especially bottled water, which is designed to give us security, health, and a sense of safety.

Accordingly, let's take a look at some of the water claims bottlers bombard us with. I can't recommend what the best bottled water product is for you, but I can greatly help you make a much better-informed decision for yourself.

"Purified bottled water such as Dasani® and Aquafina® is fraudulent" – Over the past couple of decades more than one disgruntled consumer group has attacked these types of bottled waters on the charges that these waters are nothing more than regular tap water that has been "filtered."

Is this true? Yes, absolutely 100%. Does that make these products fraudulent? Well, how about I let you decide?

IV solutions are all prepared with processed tap water. Injectable drugs and vaccines are all preparations that are based in processed tap water. Surgical instruments are cleaned with water that is processed tap water. Titanium orthopedic hip and knee joints rely on processed tap water to create inert, bacteria-free surfaces. All laboratory chemical, bacterial and viral reagent and agar preparations have as their base processed tap water. The high-purity water that is necessary to make solar panels, computer processors, memory chips, cell phones, televisions, automobile bodies and more, all require processed tap water.

The truth is that Dasani®, Aquafina®, Nestle® and other similar bottled water companies rely on complicated, well-maintained reverse osmosis

(RO) technology to create their safe, delicious, and convenient products. This is the same exact technology that is used in all the medical and high-tech applications mentioned above.

Any criticism of these great water products is not based on science and fact, but instead an uninformed, total lack of water understanding.

The actual process for making these types of bottled waters involves a precise, well-designed system and combination of known, proven and safe water treatment technologies. The typical system looks something like this:

1. Tap water or well water feed.
2. Sand/sediment filter – Removes silt, dirt, and large particles.
3. Activated carbon filter – Removes chlorine compounds to protect the RO membranes.
4. Water softener – Removes hardness to protect RO membranes from scaling.
5. RO – Removes 95%+ of mineral content, 99.9% of bacteria and other organisms.
6. Remineralization – Minerals are added to combat the flat taste of straight RO water.
7. Sanitization – A combination of ultraviolet light, ozone, and microfiltration is employed to keep the water free from bacteria prior to bottling.

What I hope you take away from this section is that the bottlers of purified tap water are ethical, technically competent, and well positioned to provide safe, convenient drinking water. Whether your need is one of convenience, safety, or emergency, this class of bottled water products is a blessing to us and should be consumed with an absolute carefree abandon with regard to safety. These are top-notch water products.

"Natural spring/sparkling water is better for the human body" – This part of the marketplace is what I consider to be the most fun of all water products. With different natural spring waters all over the world consumers of these products will be forever engaged in debates about which one is most magical, most beneficial, and most close to what nature intended for man. I call this the "fountain of youth" market. The takeaway here is that these waters are all safe and offer enough of a variety of tastes that one could establish themselves as a bottled water connoisseur reviewing each one.

It is unlikely that anyone has ever been harmed by drinking mineral/spring water. The health claims of these products are generally also the fun, folklore claims or vitality, energy, hydration, and good health. Even tap water offers these things, so if you enjoy sparkling and spring waters, enjoy, live long, and prosper.

"Alkaline water helps the body maintain an alkaline body and blood pH" – Of all the bogus, misleading, and improper water claims, this popular claim is at the very top of the list. Let me say in no uncertain terms, drinking any brand of alkaline water will do absolutely nothing to affect, normalize or otherwise stabilize the pH of your body or blood.

In writing this book I made a concerted effort to focus on water and stay away from playing doctor and nutritionist. However, as a water expert of 35 years, a biologist with years of experience as a health-care giver, with clinical time in hemodialysis, and training from very accomplished physicians, I am qualified to comment on this ridiculous alkaline water claim. So, let us have a serving of alkaline water, shall we?

- Body pH is regulated by our kidneys, lungs, and internal buffer system.[27] What we put into our stomachs, unless it is something bizarre and toxic has absolutely no effect on the pH of our blood and tissues.
- The normal pH of our stomach is 1.5–3.0.[28] This is very acidic and needs to be that way to digest complex food materials.
- Normal tap water is already alkaline and full of minerals. Tap water pH in America is required to fall in the range of 6.5–8.5 and most tap water is already alkaline.[29] Purchasing water for its supposed alkaline characteristic makes absolutely no sense.
- I have never been able to find one statistically large enough study that demonstrates any medical or health benefit from drinking alkaline water. There are, however, significant studies that show the dangers of having too much alkalinity in our bodies. These downsides include nausea, vomiting, confusion, hand tremors, muscle twitching, and tingling in the extremities or face.[30]

"Hydroge water, vitamin/electrolyte water supercharges your body" – The products covered in this section make similar claims to those listed in the last section. There are so many specialty water products on the market today that it would take a volume of books to cover all the various claims made by this class of products; however, there are some strong guidelines that do apply.

I want to be on record as saying that I take great exception to any water products that do not talk about water, but instead focus on the ben-

27 https://www.merckmanuals.com/home/hormonal-and-metabolic-disorders/acid-base-balance/overview-of-acid-base-balance#:~:text=The%20pH%20buffer%20systems%20work,(the%20corresponding%20weak%20base).

28 https://sa1s3.patientpop.com/assets/docs/18418.pdf.

29 https://www.epa.gov/sdwa/drinking-water-regulations-and-contaminants#Primary.

30 https://www.healthline.com/health/food-nutrition/alkaline-water-benefits-risks.

efits that their additives provide to our health. There is an important distinction to make here, and the consumer must be acutely in tune to what they are being asked to consume. The entire concept that some chemical chef can add ingredients to water to provide miracle benefits to customers is to me unethical and borderline unscrupulous. When a company begins adding ingredients to water and then claiming health benefits, we are no longer buying a water product but instead a nutritional product. When we purchase such products, we are allowing some unknown, unseen formulators to control our diet and health care.

Please understand, I am not condemning the nutritional supplement and products industry. Quite frankly, that industry fortifies exactly my objections to these magic ingredient water.

For your examination, examine a great product like Ensure™. It offers nutritional fortification, but nowhere in its advertising does it make claims that it is supercharged milk or supercharged anything else.

You see, my objection to fortified water companies is their claim that they are improving water, or redefining water. They are not. What they are doing is dissolving chemicals in water and telling you to drink them.

In my world, if I or my loved ones have any concerns about things like blood pH, digestion, endurance, breathing, energy level, concentration, or any other physical condition that seems to be out of kilter, we go to a doctor for guidance, not a beverage company.

Will an enhanced water help invigorate and hydrate us after a hard workday or strenuous athletic performance? Sure. But so too will regular water and a nice fat, cold orange.

My rule of thumb is to support water bottlers that boast great water. I advise staying away from beverages that claim to know more about your body than your doctor or nutritionist. If you need a nutritional supplement, your doctor can recommend the absolute best of them.

15.
Keeping our Water Clean

Concerned citizens, environmentalists and green activists are not likely to be happy with this chapter. However, it must be understood that science does not exist to support what we think, feel, or believe. Sometimes, even when we are trying to do the right thing, science will slap us in the face, contradict us, and leave us feeling totally pissed off.

When it comes to keeping our water clean, most of us are not ready to hear the reality of what science has to say. If scientific reality displeases you, this would be a good chapter for you to skip over.

I am in the throes of an ongoing love affair with water that is thirty-five years young and showing no signs of stopping. In my life I have purified water, cleaned it, filtered it, recycled it, studied it, written about it, and helped many solve water-related challenges. My love for water truly is without bounds. However, even with all that said, I am going to share an unshakable tenet of water science.

KEY WATER DROPLET – It is scientifically impossible for us to keep our water sources clean.

Are you shocked? Do not be because there is nothing humankind can do to prevent our water sources from becoming increasingly polluted. As we learned in chapter 3, contamination is the natural state of water. It thirsts for pollutants so the idea that we can keep our water sources clean honestly defies nature. But there are other reasons that it is impossible for us to keep our water sources clean.

Human activity is enemy #1 – Like it or not, you and I are the worst threat to our water. Our pollution is unstoppable, and we grow worse with every tick of the clock.

To begin, it has been estimated that Americans collectively drive a total of 2.5 trillion miles per year. In doing so we deposit countless pollutants onto our roadways. Driving leaves pollutants from rubber tire wear, oil, gasoline, and grease all over the roads. When it rains, much of this junk pours into our lakes, rivers, oceans, and streams. And what about the tars and volatile solvents that get washed out of our asphalt and concrete? Think about new pavement and how nice and shiny black it is when it is laid down. Now think about what it looks like 4 years later. You want to guess where the chemicals that once made the pavement look black have gone?

Unless we have made a conscious effort to stop driving, this pollution source will continue to increase.

Next, let us add in all the pesticides, herbicides and fertilizers that have become a staple of the American way of life. When it rains, where do you imagine this runs off to? Of course, countless tons of it washes into our water sources.

There is more! Many of today's most popular consumer products have chemicals in them that were not even on our radar 5 or 10 years ago. For instance, the chemicals collectively known as PFAS (per-fluoro-alkyl substances), which are used in products such as nonstick cookware, microwave popcorn bags, fast-food packaging, fire retardants and more, have leaked into our water sources and are highly suspect of being carcinogenic.[31] Our landfill and roadside litter are plastered with this stuff. Again, rainwater washes this into our water sources.

Let us not forget the tons of microplastics from our nuevo-tech shampoos and conditioners that make their way into our drains. Not only are these products finding their way into our water supplies, but we are also ingesting them in many ways. Remember when I mentioned that some contaminants we face today didn't even exist when our water treatment plants were built? Microplastics are a good example of such contaminants. The scariest part is that while science can detect microplastics, it cannot yet determine the effect on our bodies that they cause.[32]

Don't turn away, we aren't finished yet. Many of the pharmaceuticals and antibiotics we and our animals take get deposited directly onto the earth as feces/manure, by way of landfill and sewage treatment plants, where the endpoint again, is our water systems.

To continue, the water pollution that you and I cause both directly and indirectly is daunting. It is greater than the pollution caused by industry, which by the way, is far more regulated than you and I are. If you think I am exaggerating, please consider some of the ongoing pollution transgressions that we citizens are guilty of.

- Painting, staining and craft projects that spread paint, solvents, and pigment vapors far and wide
- Power washing of decks, siding, windows, and stone with chemicals
- Car washing

31 https://www.epa.gov/pfas/basic-information-pfas.

32 https://www.washingtonpost.com/health/youre-literally-eating-microplastics-how-you-can-cut-down-exposure-to-them/2019/10/04/22ebdfb6-e17a-11e9-8dc8-498eabc129a0_story.html.

- Burning of leaves, wood, tires, crops
- Litter, especially plastics
- Treated wood disposal, scraps, and sawdust that deposit chromium into the environment
- Wastewater from showers, washing machines and dishwashers
- Improper disposal of paint, oil, grease and other household chemicals

While we all strive to be good to the environment, the very definitions of life and civilization dictate that all living things cause pollution that winds up in our water sources. Given that the population is increasing it cannot be argued that our everyday lives won't continue to contribute to water pollution. When we stop to consider that most nations are regulated far less than the United States, the immense danger humankind represents to its own water sources looms as terrifying.

Clean energy, a Trojan Horse to drinking water – If the previous section strikes you as troublesome, this section may bother you even more. It may motivate the Green New Deal folks to send Don Corleone my way. Nonetheless, let's look at how clean energy is actually a threat to our clean water.

The casual observer looks at something like a solar farm, notices that there is no smoke, smog or chemicals coming from it, and assumes confidently that this is clean energy and is better for the environment.

In the academic world of science, however, we look, or I should say, are supposed to look, at every conceivable step associated with a process. To the dismay of the environmental well-wisher, more often than not, what such studies find is that per unit of energy produced the new, "clean" technologies are often not much cleaner than the old ones, and in some cases may actually be more polluting.

As an example, consider how the production of solar panels assaults our drinking water. While the raw material to produce semiconductor silicon is abundant, the actual production requires temperatures as high as 2,200°C. This requires great deals of energy and then produces tremendous amounts of carbon dioxide and sulfur compounds, the very things that the technology supposedly eliminates.

Additionally, solar products require metals such as silver, copper, germanium, zinc, nickel and more. Those metals can only be found by mining. So, just as it is with crude oil, the production of solar panels requires great disruption to the earth's surface. Mining causes metals to leach into the water. Furthermore, as these metals become more scarce mining will become more severe and our water sources will be even more susceptible to contamination.

The problem continues. With solar technology we also need lead, lithium, and cadmium for batteries. More mining, more fossil fuel-burning machines, and more copper for new lines of power transmission from the solar plant to the grid.

Do you still believe that so-called green technology is earth-friendly?

Not long ago the US government began force-feeding us energy-efficient CFL lightbulbs. They offer less energy consumption and by lasting longer are touted to reduce the number of glass bulbs we send to landfill. Sounds good, doesn't it?

The problem, however, is that CFLs, like coal, contain mercury. Mercury is terribly toxic to brain cells. Do you imagine that we Americans are properly disposing of our burned-out CFL bulbs? Of course, we are not. We place them into the trash with everything else and voila, mercury finds yet another way to enter our water.

Now, I am not knocking alternative sources of energy; however, if we really are sincere about protecting our water sources then we must examine without prejudice every potential contamination that both the manufacturing process, and the operating process present to us.

Industry – We could dedicate an entire library to the discussion of how industry pollutes our water. Industry contains a wide array of company profiles. Some industries criminally pollute, some industries do not pollute, and many, the majority, pollute within the limits allowed by the EPA. I am happy to report that in my own experience, the norm has been that when industry does have a spill or accidental release of contaminants, they generally report it properly, and work with the EPA to fix it.

The average citizen believes in the simple answer of zero discharge, and while it sounds like a great idea, it is not practical, and I will restate a bit differently what I have said above: The scientific laws of conservation of mass and energy, coupled with the laws of chemistry and physics tell us in no uncertain terms that this is not possible to create an industry that is totally pollution-free. Let us try to understand why!

First, from the standpoint of chemistry there are two types of reactions, and a simple graphic representation of each follows.

- **Exothermic** – These create/give off heat, or energy.

A + B ⟶ C + d + heat/energy

- **Endothermic** – These require heat/energy to be added.

A + B + heat/energy ⟶ C + d

In the equations the letters *A* and *B* represent the reaction ingredients. The letter *C* represents the desired product while the letter *d* represents the waste products generated by a given equation. To be fair, the waste products of an equation are not always harmful, and in many cases even if they are harmful, industry strives to find ways to reuse and add value to these by using them in other processes.

Still, no process is ever 100% efficient. There is always leftover chemistry in any chemical process. There is always waste, there is always a *d*. So even if industry takes waste product *d* and puts it into another chemical process, that secondary process will also have a waste component to it, we can call it "d" (d prime).

Visualize chemical reactions like those table-top perpetual motion amusement devices. You know, the ones that have a series of stainless-steel balls hanging by string harnesses from a rack.

The user pulls one ball back, lets gravity pull it down, allowing it to smack into the balls at rest. Upon being smacked, the other balls set off into motion, with the ball at the opposite end propelled away from the other balls, which now cling together. Then the ball that was projected out, succumbs to gravity, falls back toward where it was launched from and takes its turn smacking into the other balls. And so then goes the back-and-forth reaction, from side to side.

As we observe this demonstration, we see that each time the balls collide they are moved at less of a distance than the previous collision.

The total input energy of the initially moving ball is converted into a combination that can be put into a similar equation as the ones above.

Initial ball force = secondary impact force + heat loss (due to impact and friction) + sound energy

- The initial ball force represents the ingredients *A* and *B* in the chemical reactions. The ingredients are the mass of the ball *x* the potential energy of gravity, which comes from the height the first ball is held before being set free to swing.
- The secondary impact force, the balls reacting and moving is the desired result. This represents *C* in our chemical equations.

- The heat loss and sound energy are the undesired by-products that are represented by *d* in the chemical equations.

So, what happens to the sound and heat generated here? You guessed correctly if you said they got absorbed into the atmosphere. But understand, that energy is not ever lost. Both the heat and sound get converted into other natural processes—vibration, plant growth, ripples in the water, and on and on it goes.

So, we see that any reactions generate waste, and even that waste, when reused generates additional waste. At some point, the waste cannot be used any further and must be disposed of. We can get more efficient, but our processes will always generate waste and pollution.

Our goal of zero discharge is illuminated as being even more difficult when we consider that every chemical and machine used in a manufacturing process had to undergo its own manufacturing process before arriving to the manufacturing plant. That means that every ingredient in our process generated its own waste products *d* before arriving at our dock. Each of those ingredients also consumed energy and produced heat.

Remember that heat generation helps water-polluting organisms to grow, and the fuel required for machines and transportation also pollutes the environment.

The takeaway here is that industry and manufacturing are far too complex for humankind to ever create zero discharge, 100% green technologies. Can we do a better job of reducing pollution? Sure, but our expectations must be reasonable and take the entire scientific process into consideration.

EPA pollution – No conversation about the sources of water pollution can be complete without mentioning one of the largest polluters of our water, our very own EPA.

"Tom, you are crazy, the EPA exists to protect our water." Yes, this is true, but in doing so they allow you, me and industry, to pollute our very own drinking water. Let us see why this is true.

If we examine our local municipal water report, we will see that the EPA has established a maximum allowable drinking water limit for just about every toxic contamination known to us. While contaminants such as lead, mercury, aluminum, and chromium are toxic to us, the EPA does allow some, albeit small amounts of those contaminants to be present in our drinking water. The EPA has determined that if we consume those contaminants at lower than recommended concentrations, we will not suffer negative health consequences. This has two important consequences.

- It requires us to decide whether we trust these maximum allowable concentrations to be safe or if we need some type of home filtration products.
- It causes us all to redistribute these pollutants back into our environment, thereby eventually putting some of them back into our water supplies. To reiterate, every time we wash cars, take showers, run laundry, wash dishes and more, we redistribute these contaminants to places that they will one day leach back into our water supplies.

Now, let us be crystal clear. I am not accusing the EPA of poisoning our water, and you should not either. However, there is an especially important tie-in/takeaway that I have been hammering home in this chapter and the EPA regulation of water confirms this hammering in a highly visible manner.

Why do you think that the EPA does not simply set drinking water standards that prohibit all toxic substances to zero allowable concentrations?

The reason can be found in the simple chemical equations I provided earlier in the chapter. Remember, no process in life, nature or industry is 100% efficient. So too, does this hold true with water purification and filtration technologies. It is virtually impossible to reduce every drinking water contaminant to zero. All water technologies will pass some level of contaminants.

Furthermore, if the EPA set its drinking water limit to that of 18.2 megohm pure water quality, as we learned in earlier chapters, such pure water would dissolve all metal it contacted and would bring one hundred times the dissolved toxic contaminants into our taps.

Let us be clear, the EPA does an excellent job of minimizing the allowable contaminants that can go into our drinking water. However, even in minute amounts, upon considering that we humans waste, flush and dump trillions of gallons of contaminant-containing water back into our environment each year, we see more strong support for my statement that it is impossible for humankind to keep its water sources clean.

Time – Perhaps the biggest nemesis to the safety of drinking water is time. From the time manufacturing and urban population began, even before the Industrial Revolution, humankind has dumped countless tons of pollution into the earth. It has been in the environment from well before most of us were born. Water, being the universal solvent will eventually dissolve and wash those contaminants into our water supplies. It might be gradual, and it might be episodes of mass acute contamination. There is no way to tell. But for sure, those who came before us have left us a pollution time bomb, hidden in the very soil we walk, build, work, and play in. This time bomb is far too immense for us to do much about.

Mother Nature – Yes, Mother Nature, the earth, left all by herself pollutes our water supplies. What we humans just cannot seem to get through our thick skulls is that a lot of pollution is natural, and in fact necessary for the planet to exist.

To repeat, the natural state of water is to be polluted. We see a pond filled with green, stinky water infested with algae, fungi, molds, plankton and bacteria and we cringe. Yet, this is natural. Earthquakes, volcanos, forest fires, hurricanes, winds, and precipitation have been bringing dissolved contaminants into our water supplies since the dawn of the universe. The heavens above are filled with toxic gasses and metals beyond the imagination.

For sure, Mother Nature is more than capable of cleaning up her own pollution. We humans just do not occupy enough space on earth to significantly dirty the planet. We can, however, trash the environment that immediately surrounds us, and we are doing one hell of a job of that.

The problem now with Mother Nature is that we have taken her magnificent qualities of distributing seeds and pollen all over the globe and added to it all the toxic material we have invented, developed, and concocted. Sadly, what we put into the air and water routinely gets taken up into the environment by evaporation and winds. Then later the poison gets poured down upon an unsuspecting community perhaps thousands of miles away.

Mother Nature is not a major generator of pollution, but instead is a magnificent distributor of it. So, whether we like it or not, the very by-products of our human existence, very often cause Mother Nature to bring us bad outcomes.

As we depart chapter 15, I will leave with the same thought I entered with. It is scientifically impossible for humankind not to pollute its drinking water resources. In chapter 16 we will look at the good news and talk about how we must change our approach to ensuring ourselves safe, clean drinking water.

16.
Keeping our Water Safe

There is a distinct difference between keeping our water clean and keeping our water safe. While I am an advocate of conservation, good planning, and new, improved water infrastructure, I am mostly concerned with helping the consumer, the citizen, the end user to drink safe, delicious water, and do so with ample supplies.

I am steadfast in my position that there are far too many forces polluting our water for us to reverse or even level off the amount of contamination in our drinking water sources. The only way we will ever see a reduction in the pollution of our natural water resources is if the world suffers a dramatic decrease in population.

How dirty will our water supplies get? It is hard to predict; however, the data already shows that despite our worldwide efforts, our oceans, lakes, rivers, streams, ponds, and groundwater are all suffering ever-increasing levels of contamination. As we learned last chapter, some contaminants are so new we do not even know yet what their health effects are.

The good news – Please forgive my repetition if I have said this before but one of the most powerful and life-giving benefits that modern populations can rely on is that technology today can take virtually any water, no matter how contaminated, and process it into safe, delicious drinking water.

A few decades ago, some cruise ships had onboard purification systems that turned shower water, and even toilet water, back into drinking water. Yuck? Well, that is what Mother Nature does. She just does it with radiation, bacteria, oxygen, and the biological process of the environment. Recycling is here to stay and will remain a staple of our efforts to keep water safe.

Again, please understand, I am not suggesting that we abandon our water stewardship and just start turning wastewater into drinking water everywhere. What I am suggesting, however, is that if we do not start right now accepting that our water sources are going to continue to become more and more contaminated and put together a long-term plan to account for all the factors jeopardizing our water, one day soon, more and more of us will be subject to opening our tap and having poisoned water that we are not prepared to deal with.

17.

Self-Sustaining Household Water Collection and Reuse Systems.

We live in an era of doubt, heightened awareness and focus on both self-sufficiency and conservation/recycling of our resources. A frequent topic of conservation at my presentations is the idea of becoming water self-sufficient and ending our dependence on municipal water services.

The concept seems to be very logical, and in principle this is certainly a technically possible endeavor. If you think about it, a municipality is essentially a personal water system, albeit having the entire community as its family. Likewise, if a person uses well water then in essence, they are indeed their own, independent municipality.

As we explore this idea of water autonomy let us make note of some important points:

- If a well is not available, then the self-sufficient water system must have a method of collection and storage.

- When we consider designing an autonomous drinking water system, we must not only consider drinking water quality issues but also local plumbing, mechanical, and electrical codes. Also, there are EPA and local wastewater codes, and water collection/storage codes to consider.

- Comparing the cost of municipal water with the cost of operating our own autonomous system has a strong bearing on whether such a system is economically viable. Still, for some, it is not about economics but instead security.

Collection and reuse sidebar – Before we move forward, let me clarify that this chapter is an exercise in theoretical technology that is focused on the eternal question of *"Can we become totally independent from municipal water sources and become water autonomous?"*

This chapter is written to be an amusing but informational look forward into some of the challenges not readily thought about when considering our own personal quest to recycle and reclaim water. The concepts I will share are by no means meant to be a deterrent to your own quest, nor are they the only engineering perspective to consider in recycling and reuse. Instead, they are a starting point where we can begin our recycling journey. The exercise will also give you a bird's-eye view of how a water engineer goes about designing a water system.

Over the years, I have witnessed some amazing products and DIY contraptions that delight and boggle the imagination—rain barrels, gutter collection systems, and permeable driveway/street materials that allow stormwater to soak through the top of pavement and then be collected at the bottom layer for rainwater and more! It is amazing to see what people are coming up with to save water.

OK, now that we understand our caveats and base information, let us start designing our very own self-sustaining water system. We will do this with the understanding that we don't have a lake, pond or well and have to rely on rainwater and some ability to collect it.

Let's install a rain collection tank – Many of us have looked down from an airplane to discover huge open-top storage tanks sitting on farmland. Farmers catch rainwater in these tanks and use it for irrigation. So, one idea is to put out a storage tank to collect rain for our source water.

Sounds good. Let us apply some science to designing our tank.

Depending on who you ask, experts say that the average family of four in America uses between 300–400 gallons per day. For argument's sake, let us say that the average use is 350 gallons per day. This translates to 10,500 gallons/30-day month, or roughly 128,000 gallons per year.

KEY WATER DROPLET – One inch of rain falling on 1 full acre of land amounts to roughly 27,000 gallons of water.

- With this at hand, we now calculate that if we had a collection tank that was 1 acre in size, we would need to catch 4.7 inches of rain per year to have enough water for our family.
- One acre = 43,560 square feet. To visualize, our tank would have to be 209 x 209 feet. That is one big tank. Do we have enough yard for that size tank?
- A downsized 1/2-acre tank would only measure 104 x 209 feet. However, now we would need to collect 9.4 inches of rain per year to meet our family's yearly needs. Still, a big tank.
- Downsizing further, a 1/4-acre tank would measure 104 x 104 feet. Smaller, but still big. With this size tank we would need to collect 18.8 inches of rain per year to meet our family's yearly needs.

- Further downsizing, a 1/8-acre tank would measure 57 x 104 feet. With this size tank we would need to collect 37.6 inches of rain per year to meet our family's yearly needs.

We could certainly make our tank deeper to hold more water, but it is the surface area across the tank and not the depth of the tank that determines how much water we can collect. If we want to use rainwater to become water independent, we need a big tank, or at least a big surface to collect water and run it into our tank.

Rainfall across the globe varies. The Midwest may get 38 inches of rain per year. Many desert cities, like Las Vegas, can get as little as 5 inches of annual rainfall.

I think you can follow where I am going with this. To catch enough rainwater to meet our family's annual water needs, we would need a huge tank and a lot of dependable, predictable, annual rainfall. If you ever collected rainwater you would find that it must rain extremely hard and for a long time to fill a glass or jug only 1 inch high.

If our thought is perhaps to use a 500-gallon tank to collect rainwater, rest assured, our family would be for the most part without water year-round.

Conversely, if we happen to be blessed with enough land and finances to build a large storage/collection tank, we will have to depend on rain when we needed it.

Unfortunately, rainfall is not predictable, and it does not fall in the evenly spaced periods that coincide with our regular, nonstop, daily collection behavior. Our tank will have to be sized two to four times larger to account for periods of low rainfall.

So, let's suppose that we worked all through the storage issue and have a properly sized storage tank to collect all the yearly water we need. Now what?

Now we have the issue of keeping this water clean and usable. If you remember from chapter 3 and a study of the nature of water, we learned that water is unstable and without chlorine in it, which rainfall does not have, algae, fungi, plankton and other live contaminants will begin to grow in our tank. Once that happens bacteria and viruses will follow. Insects will not be far behind, and then not far behind them, critters that eat such things. Of course, now our water will face the additional contamination of animal waste.

Let us also not forget debris and pollutants that will contaminate our tank by way of the atmosphere.

If that is not trouble enough, there will no doubt be a lot of government regulations requiring us to protect the public from the tank, including physical barriers as well as demonstrable laboratory analysis to prove the water in the tank is not an environmental hazard.

Let us now assume that we are a technical/administrative genius, and our tank is legal and ready to fill.

Now we must design a pumping system with controls and a piping network to get the water to our house. Also, we will need a well-designed water purification system, secondary storage tank, repressurization system, and analytical documentation to satisfy the EPA and local regulators that we are a certifiably authorized potable water provider.

The price of all this ingenuity? Perhaps $500,000. One could certainly knock this cost down significantly by digging a pond, although no matter how deep a pond is, it can only fill as fast as surface area will allow. If we lived in a valley, the slopes around our pond could dramatically increase the collection surface area so we might be able to catch a year's worth of water in one good rainfall.

We can see that designing a workable storage vessel to become water independent requires a lot of thought, planning, engineering, and cost. Beware!

Even if we dug a pond we would still have a lot of the same issues as a tank, plus the additional challenge of how to keep water from leaking out the bottom and flooding the surrounding soil. Also, local regulations may make protecting the public from the pitfalls of a pond a lot tougher than dealing with a tank.

The takeaway here is that if your monthly water bill is a couple of hundred bucks or less, it is going to take decades to pay for a self-sufficient water system with the savings over using municipal water. For most, a self-sustaining household drinking water system is just not practical.

Let's install a grey water recirculation system – This line of thinking is interesting and fun to consider. Under this idea we would reuse as much of our water as possible thereby minimizing our costs and use of fresh municipal water. Without a doubt, systems that accomplish this can be of great value in reducing the demand placed on municipal water providers.

I want to talk briefly and sparingly about the wide range of so-called gray-water systems that are commercially available across the fruited plane. Grey water systems are designed to reuse shower, laundry, and kitchen sink water. These systems are not designed to produce drinking water, but instead for gardens, lawns, flushing toilets, perhaps car washing and other non-drinking applications.

I am not a proponent of any current technology that introduces human feces into the water reuse network, and always advise against such technology. Safe drinking water can indeed be made from toilet water, or what is called black water, but this is best left to professional water engineers and system operators.

With that being said, commercially available gray-water systems to me, are unremarkable. I say this for a few reasons.

- The best ones I have seen claim an ability to save 10,000–20,000 gallons of water per year. While this can certainly help a community save on fresh water, without being able to bathe, shower and drink recycled water, most of the recycled water here winds up down the drain anyway.
- Using this water to flush toilets presents the problem of having to plumb the recycle line into the toilets. This is going to involve breaking into walls to install new toilet feed lines and disconnect existing tap water lines. This is complicated, messy, and expensive. It also involves local building code issues.
- Also, a commercially available gray-water recycling system may not produce enough water on a given day to perform enough toilet flushes. To prevent not having enough water for toilets, a tap water line would still have to be tied in to allow enough water to flush toilets or run lawn sprinklers. Most regulatory bodies are not going to look kindly on that.
- Let's consider the economic impact that best-performing commercial gray-water reuse systems can make on our water world. Again, the best ones produce about 20,000 gallons per year for a family of four. Based on the estimates of 128,000 gallons of annual water use for such a family, commercial gray-water recycle systems are saving about 12–15% of water used. If your water bill is $200/month, one of these better gray-water systems would save you about $25 per month. At a price tag of $4,000 for one of the better systems, the payback for such an investment can be as long as 160 months.

- Finally, that 12–15% water saving is a misnomer because the concentrated waste from these systems usually requires the municipality to use more water in diluting them before they can be effectively treated and released. So, while a commercial home gray-water system will save some water on your end, because concentrated waste must be diluted to be transported and treated, a lot of the water saved with these types of systems is offset by the additional water that must be added to treat the filtered material being sent to the drain.

Let's install a true household water reuse system – I cannot see a time when home water recycle technology will treat toilet flushes and kitchen sink water. In both streams the total solid material is too dense, full of organic debris, and presents opportunities for the spread of disease. Barring a new technology being brought forward we can pretty much figure that toilet flushes and kitchen sink drain water will remain resigned to the municipal treatment plant.

All the reservations aside, the idea of reusing sink, shower, and other household water (laundry, car wash, etc.) continues to fascinate the independent spirit and inventor. It certainly intrigues me. So, what would such a potential recycle system look like? Well, it would look something like figure 14. Let us go through this step by step.

FIGURE 14

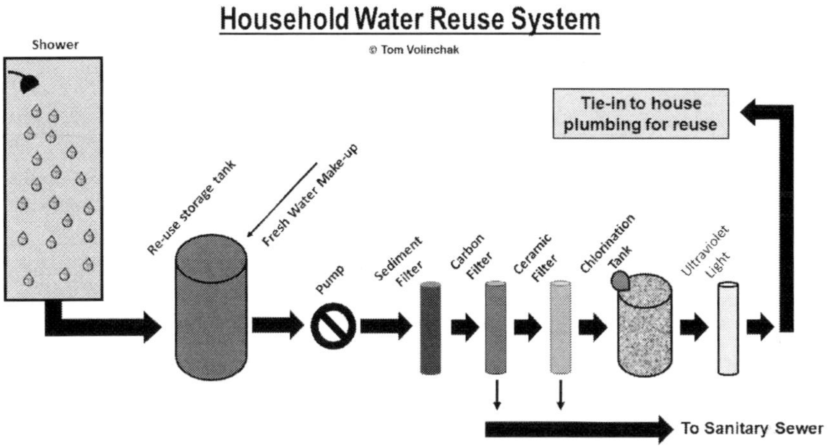

KEY WATER DROPLET – If you are wondering why there are so many pieces of equipment in this system, please harken back to chapter 4 and recall that there are 4 different types of water contaminants. Our system must have multiple components to account for all those contaminants.

Starting at the far right we show the shower. In this exercise the shower represents all the major water users in the house except for toilets and kitchen sinks. The "shower" could include rain barrel water, bathroom sinks, and even underground rainwater collection systems.

What we now must have a sobering grip on is that in our commitment to install a total reuse system, we will be introducing dangerous contaminants into our water source. In this case our water source is the reuse storage tank. This tank is essentially our water source, our tank or pond if you will. Because the filtration system is going to send 30–40% of the water it processes down the drain (we will explain shortly) we will still need to add a fresh water makeup line to keep our storage tank full to meet our daily demands. This line can be a municipal tap-water line, although if we genuinely want a self-sufficient reuse system this makeup line must come from a non-municipal source such as a rain collection system.

The third component in our system is a pump. Now that we have collected water for treatment and reuse it must be pressurized so that it can pass through filtration technology, and flow with proper force when we open our faucets. In addition to the pump, we also must add the complexity of level controls to shut the pump off when the level is high; otherwise, we will overflow the tank. We will also need a low-level switch to shut the pump off and keep it from burning up if the storage tank ever runs empty. The storage tank will also need an overflow pipe tied into the drain, and it will need a breather filter to clean the air entering the tank as the water level falls and takes in air.

After the pump, the water is filtered with some type of a sediment filter. This filter could also be called a yuck filter because it would take out things like hair, skin particles, dirt, and anything large that came off in the shower, bath or even teeth brushing. Welcome to the world of recycling.

Next in line we would use a granular activated carbon filter (GAC). We learned about GAC in chapter 8, but in this application, the carbon would be used to remove things like soaps, lotions, cosmetics, mouthwash, toothpaste and more. Note in the drawing that there is a drain line associated with this filter. This is because a carbon filter must be backwashed from time to time to clean it, remove the debris it has collected and send it to the drain.

Oops! Am I saying that all the junk the carbon filter removes is going to wind up going down the drain anyway? Yep, absolutely. Remember, we are talking about filtration of contaminants, not the destruction of them. We must not be fooled into thinking that recycling water eliminates contaminants. The contaminants are only removed from what we use, but they are still put back into the earth, by way of the drain and sanitary sewer system. Noteworthy also is that our system will put the contaminants back in a more concentrated manner.

After the GAC filter we would use some type of a microfilter or ultrafilter. Generally, a ceramic filter is a good choice because it can be cleaned in harsh acid, caustic, and steam, and used over and over. These types of filters are designed with a much smaller particle cut-off size than a sediment filter and can be designed to remove bacteria, viruses and even certain chemicals, oils, solvents, and greases.

Again, we have an oops! As you can see in the drawing the ceramic filter also requires a drain line. That is because while a filter such as this can certainly remove 99.9% or more of water contaminants, it cannot remove them from 100% of the water to be treated.

Once the contaminants have been screened out, they must be sent to the drain. The only way to send them to drain is by way of a wastewater stream. The ceramic filter creates something called reject water to accomplish this and it will send anywhere from 10–25% of the water coming into it to the drain.

The next step in our system is a chlorination tank. Even though our ceramic filter has removed 99.9% or more of all organisms, the water is not free of the threat of those organisms. Remember that contaminants like bacteria, viruses, algae, mold, and fungi are living organisms and will multiply. Even if 99.9% of these organisms have been removed from water, the remaining specimens will indeed multiply and once again pollute and contaminate the water, thereby making it undrinkable. Remember from our lessons in chapters 2 and 3 that water cannot exist in a pure state.

So then, in our system the chlorination tank allows us to inject the disinfectant, just as our municipalities do at 2–4 mg/l to kill the remaining bacteria. Many are opposed to chlorine disinfection; however, it remains hard to beat because of its high effectiveness, low cost, ready availability, and ease of distribution. I personally would never design a reuse or drinking system that did not include chlorination.

The final step in this theoretical reuse system is an ultraviolet light system (UV). Many treatment systems also use ozonation instead of or in conjunction with UV. This is yet another step in eliminating harmful organisms. What water engineers have found over decades of designing treatment systems is that no matter how effective the treatment technology, the growth of new organisms is the most difficult task in keeping water clean. If we were to really design this type of a treatment system, I promise you that this amount of attention to bacteria and viruses is appropriate.

In fact, there should be one more step in this design. I left it out because it added too much complexity to this basic drawing. The missing component would be a recirculation loop (RL). The RL would involve a small T joint at the outlet of the UV system. A pipe would be connected to it, then run to a pump, and the line would be connected back to the inlet of the ceramic ultrafilter.

The science here is that we keep the water moving through the disinfection process when we are not using water. This eliminates standing water and allows continuous disinfection of the water. Then, even when we are using water, the disinfection components of our system get repeated opportunities to kill organisms.

The final step in designing our recycle system is the tie-in, which brings our recycled water back into our home plumbing systems. This is the trickiest step in the entire process because of the regulatory issues that govern how we use and treat water.

While our system design will indeed produce safe drinking water there are no standards or regulations that I know of that allow a homeowner to treat recycled water and have government certify it as EPA-approved potable water.

This is the biggest hurdle to such systems ever becoming a reality. Right now, what motivation is there for government to approve such systems? The cost would be enormous and would involve things like approving standard designs, setting up initial and ongoing inspections, setting up, enforcing, and monitoring regular lab analysis, and of course accepting the legal liability that comes with approving such technology for the homeowner. Remember, municipalities have trained engineers and scientists on board. The homeowner generally does not have such expertise.

The other hurdle is the economics of recycling. As you can see, a true recycling system is expensive. The system I designed in this chapter would easily cost a homeowner $100,000. At that price, Theoretically, it would only save 12–15% of our water used. To be honest, in operation it would likely save even less; I would predict 4–8%.

On the surface, the enthusiasm for recycling water is universal. Because of that many commercial recycling products are already available to us, and more are likely on the way. However, if we take a deeper look into recycling, such as we have done here, the enthusiasm quickly fades.

The payback for investing in such technology, given the current low cost of tap water, would be decades. The regulations, testing and inspections add even more complexity and obstacles.

Also remember that if because of personal recycling we started using dramatically less tap water, the price we pay for that tap water would increase. Our municipalities have fixed costs that are not going to vanish just because we are using less water. The costs of staff, equipment, testing, maintenance and repair of infrastructure, and regulation are not going to be affected by citizens using less water. If the municipality must pay its fixed expenses over fewer gallons being used, then logically the price per gallon the citizen will pay for water is going to increase. With an increase in the price of water, the economics of a home recycling system get even worse.

Another note to consider, if people do ramp up to mass recycling and the price per gallon of tap water does go up, it will be the poor and economically disadvantaged who would suffer most from an increase in the price of municipal water.

I am not suggesting that we abandon the idea of personal water conservation and recycling. What I am illuminating, however, is that the viability of these things is not just as simple as saying, "Recycling is good, we must do it."

In chapter 15 an equation was provided that demonstrated how there are always waste and unseen energy costs associated with any process. We illuminated how even clean energy is not so clean and not so efficient. Here, in a conversation about water recycling we see again how there are many hidden costs and problems underlying the idea of recycling.

I inserted this chapter for a few reasons. I thought it would be a fun learning exercise to design a water system together and help you see all the things that go in to treating water.

More importantly, however, I sincerely want you to understand how crucial it is that before we embrace new technologies, we really must examine all the negative effects, hidden costs and environmental impact of the manufacturing of these technologies and bringing them into widespread use. Too many times the public is enticed to accepting a new technology, only to find out years later that the technology was burdened with hidden downsides that made it no better or even worse than what we were doing originally.

Always remember this with regard to water, and really anything scientific. Technologies that we have depended on for decades are called tried and true for a reason. A lot of science and great minds have gone into giving us what we use and enjoy today. If we are to replace them and change, these new technologies must be held to those great standards and not simply accepted because the sales pitch looks great.

Water conservation and recycling will be topics that will be spoken about more and more with each passing day. The claims of new products are sure to be mind-boggling and fantastic.

My advice for now is to stick to the simple, easy-to-operate and understand techniques. Things like using flow-restricted faucets, collecting rainwater for plants, and outside washing chores, and using low-volume toilets are inexpensive and effective ways to reduce water consumption. Leave the complicated stuff for those who can afford to take risks.

18.
A Personal Water Plan

If there was ever a topic that could bring more differing opinions and points of view than designing a personal water plan, I cannot think of what it would be.

No two residential-water treatment service companies would ever agree, no two water engineers would ever agree, and for sure no two do-it-yourself homeowners would ever agree on what the best personal water plan is. We can also add our municipal water providers to the list that would give us varying suggestions.

The challenge of writing this chapter is that citizens come in an infinite array of ages, physical abilities, mechanical know-how, scientific acumen, and economic situations. Our capabilities are not all the same. We also drink from vastly differing water supplies. What works for one may not be attainable for another. Still, every one of us deserves to have safe water, and every one of us must have a personal water plan because whether we want to face it or not, we will all be threatened with either unsafe drinking water, or a water shortage at some point in our lives.

An incredibly good model of how we should view and treat water can be found in any research hospital or university science building. In this laboratory environment there are many different uses for pure and ultrapure water. Central supply needs one level of quality for sanitizing instruments and lab glassware. Distilled or RO water is fine for this application. The chemistry labs, however, need a higher purity water to make sure trace and extraneous minerals do not interfere and give false readings to their tests. Going even one step further, the genetics and microbiology labs need a water that is highly pure from a perspective of bacteria and viruses.

So, what is their water plan? We learned earlier that high-purity water will quickly degrade and become contaminated; accordingly, it makes no sense to put in a central super-high-purity water system that delivers the highest water quality to all users when some do not need it.

The solution that all such facilities use is to install a central RO system to get a good, basic pure water to all the individual users and then let those who need higher purity water install their own personal water filters to bring the house water up to the highest standards sought. In the water industry this is called polishing technology. The filters used at the end of central distribution systems are commonly called polishers.

Now let's think about how this applies to you and me ensuring safe drinking water for ourselves.

In some very real ways, our homes and our water requirements mirror those of the medical R&D facility. We have a similar high-purity need in that our drinking water must always be impeccable. We want safe drinking water, and we know that the standards set forth in the Clean Water Drinking Water Act will assure us of that safety. At the same time, we can accept lower-quality water for watering lawns, doing laundry, and even showering. This realization provides the base logic for formulating our own water plan.

Personal water plan, step one – *Accept that our municipal water system provides us a general water quality that will not always be 100% safe.*

Personal water plan, step two – *Employ polishing water technology only on the points where we need safe drinking water.*

Our faucets and appliances will be our areas of focus. And remember, if there is a contamination alert, don't use a dishwasher or even manually wash dishes using this contaminated water.

Personal water plan, step three – *Consider it mandatory to have a substantial supply of bottled water in your home.*

In times of water crisis there will be a mad rush to purchase drinking water and stores will run out fast. Bottled water is not prohibitively expensive for the short term, and it makes sense to accumulate an emergency supply.

The amount of bottled water placed into storage is an individual choice and depends on things like the size of the family, available room for storage and personal consumption habits. However, as a starting point we can assume that the average American drinks about 3/4 of a gallon per day.[33] This estimate takes into consideration water we consume in beverages such as soda, coffee, and tea. It also considers that we get a portion of our daily water needs from the food we eat.

However, since we are trying to come up with the best possible assurance of safe water, let us err on the side of caution. Let us assume that all the 3/4 gallons per day of water we need can only come from drinking actual water. This means that a family of four will drink about 3 gallons of water per day, which makes calculating how large our emergency supply of bottled water needs to be.

33 https://www.livestrong.com/article/338496-the-average-consumption-of-water-per-day/.

It is an amazing coincidence that the average 24-pack of purified drinking water happens to contain an average of 3 gallons of water. This tells us that in the very worst-case scenario, a family of four will consume one case of bottled water per day. With a price of $3 per case being reasonable the cost of having a one-month, thirty-day emergency supply of bottled drinking water will cost only $90 per month. The same supply will only cost $45 per month for a couple, and only $27.50 per month for an individual.

The takeaway here is simple. A family of four can build a 100-day surplus of bottled water for $300 or less. The cost for smaller families is even less.

Storing 50 to 100 cases of water might be a logistics nightmare in homes where space is limited, but imagine the inconvenience and frustration we would have to endure trying to find safe drinking water during a 3-month emergency!

Whatever your personal needs are, having an emergency supply of water in any era is a good idea. Find a spot for it and store yourself some water assurance.

KEY WATER DROP – Remember that bottled water is not chlorinated and once opened is prone to growing bacteria and other organisms. God forbid we ever must rely on bottled water for an extended time, let's not keep an un-chilled open bottle of water any longer than 24 hours. Seal and refrigerate opened bottles and feel safe to keep them for 48 hours. I recommend staying away from 5-gallon bottles which, once opened, usually sit for longer periods of time than a smaller, single-use bottle. Finally, store your bottles in a cooler place out of direct sunlight. Sunlight can degrade some plastics and contaminate your emergency supply. Sunlight also stimulates the growth of many harmful organisms.

Personal water plan, step four – *Carefully choose the proper polishing, point-of-use water filter.* As we mentioned previously the majority of faucet and appliance water filters are GAC filters, which are fine for removing undesirable taste and odor components as well as chlorine compounds, however, do extraordinarily little, if anything, for toxic metals and harmful living organisms. If you have one of these types of filters and your municipality issues a boil alert or other toxic contaminant alert, it would be wise to replace those filters once the bad water notice is rescinded. Such filters are prone to biological contamination.

Look instead for filters that also promise heavy metal removal and elimination of microorganisms. If you cannot find such specialty filters it would be best not to depend on the common point-of-use drinking outlets during the water contamination alert.

At the present time reverse osmosis technology remains the best overall purification technology for guaranteeing safe drinking water. When operated properly it removes virtually all biological contaminants, and 95%+ of toxic metals (copper, lead and arsenic for instance).

If the goal is to have a drinking water system in place if, God forbid we exhausted our emergency bottled water supply, an RO system is the technology to use.

However, as we discussed previously there are some caveats with RO systems. RO systems are slow to produce water so generally a storage tank and a pressure pump are required to provide the fast glass fill rates that we are used to. The storage tank and the lack of chlorine in RO water are challenges we must address if we choose this technology. Professionally designed, this technology can indeed supply normal faucet flow rates, but it will require proper bacteria-free design, and a system large enough to handle our demands.

At this point in the book, I am confident that you can now ask vendors and product representatives the tough questions to make sure that they provide the proper technology and eliminate these secondary biological contamination issues.

But make no mistake, I believe it to be crucial that point-of-use polishing filtration technology, and a backup technology like RO, be installed and ready to perform in your home should the local municipality have a contamination issue.

Personal water plan, step five – *Flushing and sanitization are crucial after the crisis.*

As a rule, after a municipality corrects whatever it was that caused the water alert, they send out notices to users to perform an adequate flush to drain, and then resume using the water as normal.

Unfortunately, we are never told exactly how many gallons to flush, and we are seldom given any guidelines or specific tests we can do to make sure the contaminant is no longer in our system. Understand, this is not necessarily a criticism because, as we have learned, the complexity of water behavior and water analysis makes managing contaminants next to impossible for the average homeowner.

Municipalities are generally good about coming to your home upon request and doing a test to make sure the contaminant is gone. Unfortunately, a negative test for contamination on Monday afternoon does not necessarily guarantee that that contaminant is not in your water that same evening.

The problem with municipal line contamination is that there are countless thousands of miles of pipe network, with pumps, tanks, fittings, and valves all along the way. Contamination can hide in dirt clumps, air pockets and areas where water use is low, only to break loose at any random time and make their way to your tap.

While I have advised against installing whole-house purification systems, I am a proponent of whole-house failsafe purification technology. Specifically, I recommend that a 20–50-micron filter, sufficiently large enough to allow full flow to your home, be installed at the very inlet when your water first comes into your home. A pressure gauge on the inlet and outlet will inform you when the filter becomes plugged.

When the pressure difference approaches 12–15 pounds, it is time to change the filter. In general, these filters cost less than $35 and can conveniently and easily be replaced by the user two to four times per year. Any qualified vendor can install such a filter, as can a decent do-it-yourselfer.

Where these filters can be a lifesaver is when your municipality experiences a line break. Line breaks cause mud and caked-on contaminants from inside the pipes to be released into the piping network. Once repaired and you open your home fixtures to rinse, you are allowing whatever is in the mud and crud to pass through your home, putting you and your loved ones at risk. And remember, this contaminated material can get caught up in your pipes and gradually be released into your drinking water over a long period of time. No thanks.

After a line break, your filter may clog up daily for a few days. If it does, be grateful because otherwise you might be drinking this stuff. Also remember, water that is flowing out of a fixture always appears to be clear and clean. Once collected in a glass, however, we often see color or particles.

A good household filter on the incoming water will trap all that mess.

In general, a polypropylene filter housing called a 20" Big Blue is readily available from local water vendors and even plumbing supply houses. The filter can be wall mounted and holds a 4-inch x 20-inch standardized filter. This is a part of my personal water plan and you should investigate it as well.

Another great use of such a filter housing is to disinfect and/or clean your entire household plumbing. It works like this: After a contamination alert is over, the filter housing can be filled with detergent, chlorine bleach or even a commercially available sanitizing agent like peracetic acid.[34] Once the filter housing is loaded and screwed back on, opening all your faucets will allow the sanitizing agent to run through your plumbing. The use of a common test kit for the chemical will alert you as to when it has made its way to your faucet. Once you get a positive test, shut the faucet(s) and let the chemical set for the specified period. Finally, after sufficient contact time open all your faucets and rinse the chemical to drain until testing shows that the chemical is all gone. At this point your house plumbing will have been sanitized.

This procedure can be a DIY operation; however, care must be taken to make sure that all the chemical is rinsed completely before drinking from the tap again. Generally, a local residential water treatment specialist can help with a proper sanitization and rinsing. It is also important to check with local codes prior to installing a whole-house filter.

In addition, a whole-house ultraviolet light system installed before the filter mentioned above accomplishes wonders at minimizing and virtually eliminating bacterial and viral contamination. These units are exceptionally low maintenance and generally only require a bulb replacement every one to three years.

The steps mentioned in this section will give your family a higher level of water security and by not changing the overall chemistry of the tap water, do not require intense monitoring and maintenance.

KEY WATER DROP – Never install an ultraviolet light system without having a filter cartridge after the UV. If the UV bulb or

34 https://www.cdc.gov/infectioncontrol/guidelines/disinfection/sterilization/peracetic-acid.html.

containing tube breaks, the filter will catch the resulting glass particles and prevent them from entering your plumbing network.

A personal water plan for the future

It is inevitable! The time is going to come when local and regional governments accept that they are no longer able to provide water that passes the standards of the Clean Drinking Water Act 100% of the time.

It was estimated in 2018 that it will take a minimum of $1 trillion to bring America's aging infrastructure up to standards.[35] Some estimates run as high as $3 trillion. There simply are no funds budgeted for this magnitude of water funding and it does not look as if there is any plan for it soon.

Furthermore, contamination sources, population growth, and infrastructure failures will continue to spiral. At some point it will become clear to leaders that the new path to providing safe water for everyone is to facilitate homeowner filtration in conjunction with the municipal supply. Such an approach will be far more practical, economical, and executable than trying to overhaul our total infrastructure.

I can foresee a new era of tax credits for home water technology and even government assistance and educational materials to help citizens choose their individual home filtration technology.

I believe that all new home construction should employ a different plumbing design to plan for this coming trend. I suggest plumbing new homes with two separate piping systems.

- One system supplies non-drinking faucets. It would include things like outdoor hose faucets, washers, toilets, baths, and showers.

---

35 https://www.nrdc.org/resources/go back-well-states-and-federal-government-are-neglecting-key-funding-source-water.

- The second system would supply only drinking outlets. It would include sinks, dishwashers, and appliances.

Such a home would relieve the homeowner from shutdown concerns regardless of what pollutants the municipality might send down the pipeline. The bad water would go directly to noncritical applications, while all drinking outlets would have safe water delivered courtesy of a central "high safety" drinking system.

I want to make an important distinction here lest my readers think I just contradicted myself. Earlier in the book, perhaps several times I advocated against the consumer putting in a central water purification system.

Under current conditions, I am still against central systems for all, save for seasoned water experts and engineers. The complexity, regulatory issues, and potential health risks given current home plumbing, as mentioned before, present too many risks for the average homeowner.

In the future, however, a home should be designed with a proper safe water network that includes a recirculation/sanitization loop, high-purity water piping, and space for a properly engineered drinking water system.

I believe that we should already be thinking of these smart water designs, and in the next chapter I share my beliefs regarding our need for a national water plan and a national water czar. This is the kind of forward planning we must engage in now if we want to keep safe water as the norm rather than the exception.

It is my hope that after reading this chapter, you too, begin putting together your own long-term water plan.

19.
Climate Change, Conservation and Drought

The primary focus in this book as stated is safe water for everyone. The topics of this chapter are of great global importance, and they certainly threaten our water. However, if we are being honest, there is little the average citizen can do to combat any of these things.

I understand and encourage the "if we all pitch in together" attitude. We all should do whatever we can to help the global water situation. However, these are immensely complicated and difficult challenges that require financial, engineering, and technical approaches that are far beyond what you and I as individuals can do.

In our continuing water journey, we must accept that so far, man has been ineffective at predicting hurricanes, earthquakes, tornadoes, and volcanos, and at fixing drought, and reversing diminishing water supplies. Despite all our efforts, you and I still face growing contamination problems. Making things harder, we have jobs, families and other responsibilities and only have limited time to devote to changing the world.

Still, no book on water would hold water if it did not at least mention the issues of climate change, conservation, and drought. Accordingly, this short chapter will immerse us in these topics. Water puns intended!

Climate change – The climate has always changed on earth and it is certainly changing as we speak. There can be no argument to this. The important matter, however, is how you and I deal with this in our personal water journey.

It might surprise some that I have confined this complicated topic to just a blurb instead of devoting a full chapter to it. There are good reasons for this. The first reason is that the topic itself is very polarizing and I have witnessed arguments in which long-time friends turned into enemies over it.

The second reason is that the topic is so diverse and immense that there is still great disagreement among scientists who are working in this field. This confounds our ability to make heads or tails of what can or cannot be done. It appears that the lesson we are to learn from climate change study is that we should expect less rain, more drought and increasingly severe weather. That accepted, again I ask, what can you or I do about this other than become smarter and better prepared when it comes to our own personal water needs and challenges?

One of the frustrating things about studies on climate change is that often the different data and observations contradict each other, presenting even more confusion.

For instance, if we look at things like the USGS drought monitor, and rainfall statistics, NOA and NASA Earth Observatory weather data, we find that right now, when supposedly climate change is causing drought, many places across the globe are experiencing all-time high rainfall.[36]

If the climate change scientist is claiming that both drought and excessive rainfall are examples of anthropogenic (man-made) climate change then following along with this campaign is even more confusing. If the proponents of this theory cannot attach defined quantities, time frame, actual expectations, and future event locations to said theory, how can we even know if our efforts are working?

As a nation, in my opinion, we are currently doing a scant minimum to alleviate the water contamination, delivery, and shortages we are facing. Not to beat a dead horse, but the big picture still comes down to you and me making smart, visionary decisions to ensure, as best we can, safe water for ourselves. Let's leave climate change discussions for those who like to argue.

Conservation – There is a crucial caveat to remember when it comes to the conservation of water. In times of short supply rationing and conservation are surefire ways to guarantee that everyone gets a fair share of the available water. It can also prolong the depletion of an existing supply. What conservation cannot do is compensate for too many consumers drawing from a given, fixed water source.

Two people can drink plentifully from a 5-gallon water cooler bottle. Two hundred cannot. Conservation will never overcome our purposeful overpopulation and aggressive urban development of areas with fixed and finite supplies of drinking water.

Conservation is a crucial and important tool in managing our water. However, it is a temporary bandage and not a solution. Personal water conservation is very ineffective at solving water depletion and shortages. Our goal should always be plentiful water for everyone. We should feel

36 https://earthobservatory.nasa.gov/images/147006/excessive-monsoon-rains-flood-asia.

carefree and guiltless about splashing about. We are smart enough to make this happen, but sadly, for reasons unbeknownst to me, we do not.

One of the frustrations with setting water conservation standards is that much of the world really has no need for them. If you are a resident of the Midwest Great Lakes area, the idea of water conservation is borderline idiotic. In addition to having trillions of gallons of fresh lake water, countless streams, rivers and smaller lakes and ponds, this part of the country has more annual precipitation than they know what to do with.

Conversely, if one lives in an area of infrequent rainfall and depleted water resources, conservation is all too often a band-aid that does not provide a long-term fix.

Things like low-volume toilets, flow-restricted fixtures and collection barrels on home downspouts help, but in all honesty in most communities the added water consumption that comes with population growth and increased real estate development offsets even the best personal conservation efforts. I have yet to see a single community where conservation efforts have stopped or slowed the sinking levels of local lakes, reservoirs, and wells, and believe personal water conservation to be impotent when it comes to improving our water situation.

Follow along with the logic.

- The average family of four uses 350 gallons of water per day, roughly 90 gallons/person.
- With efficient toilets and faucets now standard, additional household conservation of 20% of water, or 70 gallons, would be extremely difficult to achieve, but for an exercise let us say we can do it. Remember now that not all households have four members. Smaller households will save even less water.

- Every new resident is going to use 90 gallons of water, essentially eliminating what the "typical" family of four saved.

- Likewise, visitors and guests of a community are going to add to water consumption that is not accounted for by the family-of-four conservation prediction. So, we lose more water.

- The USDA[37] verifies that 80% of our water consumption is agricultural use, and as high as 90% in some western states. Industrial use is perhaps 10%, leaving 10% of water used in America to be used by residents.

Now think hard and long on this: If every citizen were able to reduce their consumption by 20%, it would only result in an overall water savings of (20% individual water savings) x (10% of total water consumption) = 2% overall water savings. The truth is that other than in places where water shortages are critical, individual water conservation has little effect on overall water supply volume.

• Making matters worse, while you and I are being made to ration water and feel guilty for using it, the nation's utilities are wasting and losing trillions of gallons of drinking water through leaks in our aging and crumbling water infrastructure.

Research estimates that the US wastes 2.1 trillion gallons per year because of leaking, aging, neglected pipes.[38] Do you still believe conserving household water is the real solution to water scarcity? The reality is that we are being asked to conserve water because our national leaders have murdered our water infrastructure and have done little to protect our water supplies. Now, there are places in the nation, like Las Vegas

---

https://www.ers.usda.gov/topics/farm-practices-management/irrigation-water-use/#:~:-text=Agriculture%20is%20a%20major%20user,percent%20in%20many%20Western%20States.

38 https://www.npr.org/2014/10/29/359875321/as-infrastructure-crumbles-trillions-of-gallons-of-water-lost.

and El Paso, that have remained ahead of the curve and do a great job of managing their water supplies. However, on the national front our water efforts have been largely impotent.

Stormwater! Now we're talking – If there is one area of water science that can dramatically help with water shortage it is in the area of capturing stormwater. Stormwater of course, is defined as the voluminous amount of water that runs into our sewers every time it rains. During a storm, streets turn into rivers, building roofs and awnings turn into waterfalls, and parking lots turn into lakes. Everywhere we look, water flows like it were the apocalypse.

In water-scarce areas, collecting this water is one activity that can truly make a difference.

Unfortunately, the impact of collecting rainwater/stormwater for you and me individually, or even for business, is seldom significant. The reason is that collecting a significant amount of water requires an immense collecting area. Most of us do not own enough land to create a collecting surface large enough to impact our total water gallonage needs.

The good news is that by considering the totality of our streets, slopes, highways and parking lots, the immense water collecting area does exist. The challenge is now, how do we get the water off all these hard surfaces? Do we use a huge Shop Vac®?

Municipalities already have sewer systems in place to collect stormwater. What very few have, however, are the screening, collection, and filtration systems to process and turn this water into usable, potable drinking water.

Ideally, we should have the national funding and planning in place for this type of water advancement, but sadly we do not. The expense of such infrastructure investment is usually out of reach for most municipalities that have scant tax dollars to spare. In the following chapter 20 we will talk about our need to have a national water plan wherein financing for these kinds of projects would be key.

One interesting and emerging technology is something called permeable pavement. This is a pavement material that allows water to pass through it, top to bottom instead of across the top as with conventional concrete and asphalt. Figure 15 gives us a glimpse of what this technology might look like.

The concept is that rainwater would percolate downward through the pavement into a sediment/dirt filter, which would send the silt to standard waste treatment via the sewer, then finally percolate into a collecting chamber that finally sends the rainwater to a treatment plant where it can be processed into potable/drinking water.

This type of a system is likely not cost-effective for the individual, but could very well be for smaller, water-starved communities. A visual representation of such a system can be seen below.

FIGURE 15

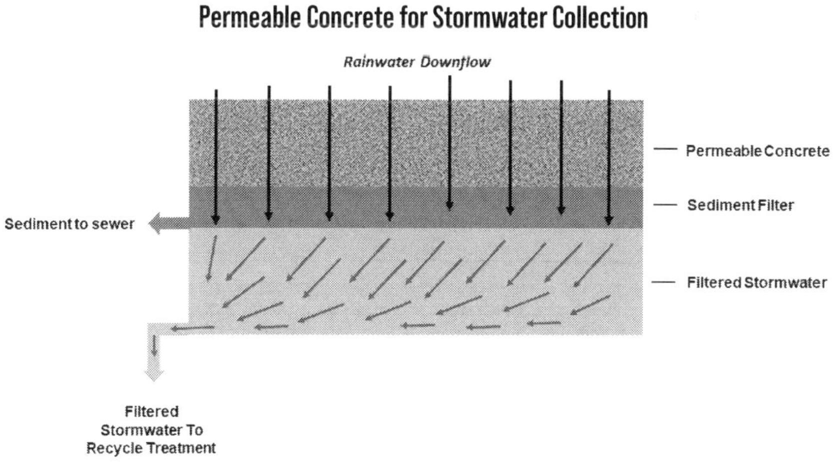

Drought – Drought is a terrible condition that can devastate communities, businesses, and civilization itself. While it is not popular to say, some drought is predictable and when it is, we have nobody to blame for its devastation but ourselves.

Case in point, we often hear about the horrible drought that devastates California, especially the Southern portion. However, if we examine the annual California rainfall patterns over the past 100-plus years what we find is that the rainfall has been relatively stable and predictable, featuring regularly varying periods of both abundance and scarcity.[39]

The perception that California does not get rain is false. The rain may be more intermittent than many other places in America, but between regular rainfall and water from the spring snowmelt California has more than enough water to provide it abundantly to all residents who need it.

Over the past decade, on more than one occasion California has dumped billions of gallons of fresh water out of its reservoir, allowing it to flow to the ocean.[40] Can you imagine this insanity? A state cursed with drinking water shortages and out-of-control forest fires routinely dumps its reservoir into the drain.

Why would California do such a stupid thing? It is simple. The rain and annual snowmelt provide more fresh water than the reservoirs can hold. So, instead of having the vision over the years to build more water storage, these leaders dump their bounty of water, and then blame nonsensical things like climate change and drought on their lack of water.

I would never marginalize the impact of drought and climate change as it applies to our global water situation. At the same time, it is inexcusable to use these phenomena as scapegoats for our total incom-

39 Data provided by NOAA's National Climatic Data Center (NCDC) and taken at station #USW00093193.

40 https://www.forbes.com/sites/ashoka/2015/04/15/why-does-california-let-billions-of-gallons-of-fresh-water-flow-straight-into-the-ocean/?sh=687ceb5b517c.

petence and lack of water planning. As I travel the nation the lack of substantial water infrastructure projects strikes me as both disturbing and disappointing.

The following quote from the poem "Rime of the Ancient Mariner," written by the English poet Samuel Taylor Coleridge in 1789 should remind us of just how badly man has done with harvesting technology to bring safe water to all people.

"Water, water, everywhere, Nor any drop to drink."

When you analyze that 71% of our planet is covered by water, which most estimate at 326,000,000,000,000,000,000 gallons (326 quintillion gallons, or 360 million trillion gallons), the thought that there are people getting sick and dying because of a lack of water, in my opinion, is one of humankind's biggest and most horrific failures.

How horrific? The earth provides 47 billion gallons of water for every living human being. How can it be, other than by our own failure, that any human being does not have safe water?

I am never one to criticize and not offer solutions. So. I will offer up chapter 20 that is entitled "A National Water Plan." It highlights the many ways America can do a lot better job of providing safe water for all of us.

20.
Our National Water Plan

If I could describe the leadership, vision, care, and innovation that our government has applied to our water over the past 50 years, I would use the phrase fatally inadequate. Oh, we still have it better than most nations and our local water suppliers still provide us with high-quality, safe drinking water. But things are changing. The threats to our water are increasing in both frequency and severity with every tick of the clock. I am here to lament that we are woefully unprepared for what is coming.

In recent times we have had to face the grim realities of water crises that have included drought, contamination, water shortages, and climate change. There are reports daily of municipal tap waters that violate the Clean Water Act (CWA). Sadly, the horror in Flint, Michigan, is becoming more the norm than the exception. Add to this flammable water associated with fracking, aging infrastructure, and concerns over having enough trained water professionals to operate our systems, and our fears about the safety of our water can no longer be marginalized.

As mentioned previously, in 2015 our own EPA was responsible for accidentally introducing 3 million gallons of contaminated mining water into the Animus River, a tributary of the great Colorado River. The contaminants included metals such as zinc, copper, arsenic, beryllium and more. Even those charged with protecting our water have been guilty of contaminating it.

Another critical concern is that many of our existing treatment plants utilize older technology and lack the sophisticated analytical instrumentation required to recognize and treat contaminants that did not even exist when the plants were constructed. Lead, fluoride, and hexavalent chrome continue to challenge us while new contaminants such as PFAS and microplastics are just now rearing their problematic heads. God only knows what new contaminants await us down the road.

Sadly, our potential water woes do not end here! An issue that we seldom talk about is the physical security of our water supplies. With increasing threats of terrorism, our need for new, heightened security programs that better address potential violent assaults on our water must be given a high priority.

As concerned as we are, as passionate as we are about our water, our concerns and activism are not enough to keep our water sources safe and clean. We all claim to be passionate promoters of clean water, but are we really? It seems as if the only time we Americans really heed the call to arms for water, is when it affects us directly or when there is an emergency that hits the news. We certainly lament the water woes of others, but as long as we, ourselves, open our own faucet and have clean water come out, water is really the furthest thing from our minds.

Can you imagine if the airlines performed maintenance and safety checks in the same manner America manages its water resources? Can you imagine going to the doctor or following a dietary and exercise regimen only when there was a severe problem? Of course not, yet with water, one of the most vital ingredients of life, we continue to behave haphazardly.

I have made a case before and make it strongly again, that if America is to enjoy safe, clean water going into the future, we must develop an aggressive, expansive, bold, forward-looking, and unified national water plan.

We must then appoint a water czar, implement financing, and then go about creating the structure and process to make the plan a reality. Without such an effort, we face threats that can destroy our health, our economy, and honestly, our very existence and way of life.

It is a national tragedy that we Americans do not recognize and understand the full impact water has on our existence. Our narrow view of water centers around our thirst for drinking water, and neglects to recognize the economic, technological, health and educational benefits that water brings to humanity.

In the era of Elwood Mead, circa 1936, the single largest water project in the history of humankind was undertaken and completed. I am speaking of course of the Lake Mead/Boulder Dem/Boulder Hydroelectricity project. This project gives us a magnificent peek into just how beneficial major water projects and planning can be.

The man-made lake itself is 112 miles long, 532 feet deep at its deepest, and when full contains roughly 9.5 trillion gallons.[41]

However, the physical and engineering features are only a part of this impressive feat.

The real jaw-dropping reality of this project is that it provides America's largest national recreation area, with countless fishing, boating, swimming, camping, and exploring opportunities, spread over 1.5 million acres of mountains, valleys, and canyons.

41 https://www.nps.gov/lake/learn/nature/overview-of-lake-mead.htm.

In addition, this water project has brought electricity, industry, and commerce to over 20 million people in the Nevada/Arizona/California area. The momentum of the real estate boom created by this water project shows no sign of slowing, a full 84 years after its completion.

So, you see, there is far, far more benefit to investing in water than just providing drinking water to thirsty residents. Water provides life to our bodies, and to our hopes, dreams, ambitions, and work.

Studies done by Water.Org demonstrate that each and every dollar invested in water infrastructure and technology generates $4 of economic benefit. Imagine that! Investing in water will get us clean water and jobs and economic prosperity. So, explain to me again why America does not have a national water plan (NWP)?

Here are some of the more important things an NWP can do for us. In my opinion these are things we should have been working on all along, and now must play catch up.

- **Add water education to elementary school curricula** – As a nation we are poorly educated in the science of water. With water challenges growing in number and severity we must be a society that has a better understanding of water, and its importance to us all. Elementary school is the crucial starting point for this to happen. Our children need to learn about water.
- **Add water education to secondary school curricula** – The need for trained water professionals and technology developers will continue to grow as our water challenges continue to grow. Our massive thirst for clean water, sustainability, recycling, and waste minimization is unquenchable and has already created a job market with more jobs than candidates to fill the positions. Our colleges and technical schools simply must begin to offer programs to educate and certify the new water professionals our nation is calling for.

It was estimated in 2018 the global water market was roughly $700 billion, and it has been steadily growing ever since.[42] Jobs as operators, chemists, biologists, IT specialists, sales engineers, laboratory/RD specialists, product developers and more are there for the taking. Most sources believe that there will be a shortage of trained professionals in the coming years. If ever there was a call to water leadership, it is in this area.

- **Provide home water treatment guidelines** – While it is not the government's job to tell citizens what filtration products to use or not use, as we speak, there is no national standard or handbook for water products manufacturers, technology providers and DIY projects. There simply is little to no regulation of consumer water products. Consumers are overwhelmed, confused, and inundated with products and claims that are simply too difficult to digest given currently available water information. A proper national water plan can change this.

- **Implementing a plan for infrastructure upgrades and maintenance** – By all estimates we do not currently have the funds to bring our aging water infrastructure and treatment plants into the 21st century. Perhaps it will take 20 years to do so, but right now we really have no plan and no set of standards to get all of America on the same page. Our systems are a mess, and our approach is haphazard. If we do not become focused immediately, things are going to continue to get worse and our tap water likewise will become less safe.

We often speak about these impending dangers but then, as is human nature, as long as our tap water is still safe, we move on to life's daily activities. However, a sobering report by Giuliani Partners in 2013 points out just how dangerous our water infrastructure is. They illu-

42 https://www.hydrofinity.com/blog/the-growing-water-industry-where-the-jobs-are.

minate that 24% of our pipers are over 50 years old with some being installed circa 1860.[43] They document that we suffer as many as 75,000 sanitary sewer overflows per year, dumping raw sewage into our water and environment. Make no mistake we are at critical levels for needing infrastructure overhaul.

- **Create tax credits for consumer water investments** – This would be a productive way to encourage conservation, reuse, personal water security and the purchase of energy-efficient products like ice makers and hot water tanks. We have a variety of programs in place to stimulate our investment in clean energy and electrical improvements. We should also benefit from investing in our water.

- **Construction of the great American water pipeline** – The only thing sadder than seeing our forests, homes and businesses destroyed by wildfires is the realization that America has an excessive abundance of available fire-quenching water that could have minimized the vast majority of these tragedies.

  Think about this: the Keystone Pipeline, a noble project, reportedly cost upward of $5.2 billion, and spans 3,000 miles with another 1,000 miles scheduled for construction.[44] Can somebody tell me why we can transport oil 4,000 miles by pipeline, but like obtuse dunces we don't transport life-giving water from our lakes and oceans to areas that desperately need it?

  It is surreal to see California burn while only a few hundred miles away is an ocean with such water abundance that wildfires can and should become a thing of the past in our lifetime.

43 http://servicelinepartner.com/wp-content/uploads/2018/03/The-Hidden-Danger-of-Failing-Water-Infrastructure.pdf.

44 https://www.nrdc.org/stories/what-keystone-pipeline.

Accuweather Inc. reports that in 2020 the damage caused by wildfires could cause as much as $150 million in damage; horrible, irreparable damage.[45] Can you imagine how different life would be out west if we had invested the same $5.2 billion for a water pipeline that we did for the Keystone oil pipeline?

A network of pressurized water pipes would allow rapid and substantial water response when fires break out. The wilderness lands, trees, property, and lives that would be saved by such a project must leave us asking why we do not have such a project already well underway? How much suffering and devastation must we suffer with life-saving water just waiting, a stone's throw away?

Our need for water leadership is so intense that I believe it beyond the ability of adjectives to describe.

And there is more good news to building a national water pipeline. Remember, Water.Org outlines that every dollar spent on water infrastructure results in the creation of $4 of new economic growth. So, in addition to saving property, businesses and lives, a project like this would create jobs, purchase goods and expand our economy immensely.

Also remember that unlike an oil pipeline, a leak in a water pipeline is not going to destroy the environment. It is also reasonable to believe that the cost per foot of a water pipeline would be substantially less than the cost of oil pipelines.

Increased monitoring and security of our water supplies – We do not like to talk about it, but our water distribution networks represent a great opportunity for deliberate contamination and terrorist-type attacks. A contaminated water supply is far more dangerous than any bomb.

---

45 Mark Puleo, Accuweather Digital Journalist, September 14, 2020, 13wmaz.com.

Through better analytical/monitoring, instrumentation, and security measures we can and absolutely must make our water supplies and systems more secure.

Once a water source or supply has been sabotaged, in addition to the horrible outcomes it could cause to lives and health, it could also irreparably contaminate the supply for many years to come. This is another important area that cries out for National Water Leadership.

Creation of a national water czar – There are many ideas that can provide water benefits to the American people. However, without a formal office, charter, funding, and an actual leader to organize, solicit and implement new ideas we will continue to splash in the water instead of making waves in the water.

The potential upsides of implementing a national water plan and appointing a water czar are boundless. The downsides of not implementing such a plan are too frightening to think about.

21.
Water Tidbits and Musings

I decided to create a fun, hodgepodge chapter where I can share some water trivia as well as a few crazy things that have happened along my watery journey of life. After all my hard work I thought I deserved a diversion. I hope you enjoy this short ride.

Blowing up fish with water – On the lighthearted side of water, even though this was not so funny for the fish, I want to share a story from my early days as a budding water systems sales engineer. It was a lesson I learned about water that at that point in time I never knew about.

On a cold, dark, icy morning in 1986, on the shores of Lake Erie in Cleveland, Ohio, I was sitting in a diner with an older colleague eating breakfast and literally shivering from the blinding blizzard that was being heaped upon us. Halfway through my pancakes, my colleague came back from the pay phone looking white as a ghost. (Yes, people, once upon a time, there were no such things as cell phones.) My friend exclaimed, "I have to go, all the fish are dead!" I initially laughed but my friend quickly scolded me: "@#$%^, Tom, this isn't funny, all the fish in the aquarium I did work for died, and they are demanding to see me. They are blaming our water system for the dead fish."

Now, I bet you are asking, how can water blow up fish? Well, please, read on.

As it turned out, my colleague was not very well versed in biology, physics, and chemistry, and he sold a high-purity deionized water system to a commercial aquarium.

The problem you see had to do with something we discussed in chapter 6 called osmotic pressure. In a nutshell here is what happened:

The pure water in the aquariums created a situation in which across the fish body tissue salts passed from the concentrated area (inside the fish), to a less concentrated area in the tank. At the same time, the pure water also obeyed the laws of osmosis and passed from a low-concentrated area (in the tank) to a more concentrated area (inside the fish). The force of osmosis in this case caused the fish literally blow out like inflated balloons.

While the fish did not actually explode, they puffed out and expanded to the point where they looked like cartoon fish. Sadly, about 300 fish perished, and my friend never forgave himself for his mistake.

There are two lessons here. The first is to never put deionized/high-purity water into your aquariums. The second lesson, which is something water scientists are working on every day, is that osmotic pressure caused by the tension between highly concentrated solutions and dilute solutions can be used to generate force, energy and drive certain chemical/mechanical processes. The force can be enormously powerful, and very dangerous.

Water for radiators and humidifiers – Manufacturers commonly recommend distilled water for things like radiators and humidifiers. But is this a good recommendation?

Contrary to popular belief, distilled water is not free of minerals. By scientific convention distilled water has a conductivity of 200,000 ohms. This translates to a total dissolved solids (TDS) of about 2.5 ppm. Most distilled water starts with tap water, so the bulk of those contaminants reflect those in the tap water. If a water softener was not used before the still, most of the ions in distilled water would likely be calcium, the element associated with scale in pipes and radiators.

We learned earlier that calcium is a weird metal because as the temperature increases, unlike other contaminants, the solubility decreases. In other words, the hotter the water, the more scale buildup will take place. So, distilled water is indeed better than tap water, but it can still, over time, contribute to scale buildup.

If you have an older vehicle and replace water in your radiator often, then even distilled water may cause scale. Admittedly it will take a lot longer to do it, but we are learning about water, so these technical sidebars are great theoretical lessons. In a car radiator, the effect is probably not a big deal; however, in a critical industrial setting, like for instance a high-pressure boiler in a nuclear power plant, contamination of as little as 2.5 ppm calcium could cause hundreds of thousands of dollars of damage, repairs and maintenance costs.

So, what kind of water is best for these types of applications?

If you can find commercially processed RO water, purchase it because it will be superior to distilled. Here is why. Reverse osmosis systems, like your radiator are subject to calcium scaling. For this reason, commercial RO systems are pretreated with a water softener. Depending on the incoming feed water quality an RO system typically produces water that is calcium-free with a TDS of anywhere from 1–10 ppm.

Most of the TDS in RO water will be sodium. The benefit here is that sodium does not readily form scale, it stays in solution and is kind to metal surfaces, even providing a small detergent effect to help keep surfaces clean.

The best water you can put in your radiator is either totally pure deionized water or softened RO water. Both waters are devoid of calcium and magnesium, which are the hardness elements that ruin radiators. For humidifiers, RO water is best.

Now to you astute readers who say to me, "But, Tom, earlier in the book you told us that pure/deionized water would corrode and dissolve metals, like those used to make radiators." This objection would be 100% correct if it were not for one issue. Remember that we mix antifreeze with water inside our radiators. In this case the corrosive nature of pure water would be pacified by the ingredients in the antifreeze.

Cruise ship water – Three years ago my lovely wife, Cathy, addicted me to Norwegian Cruise Line cruises. So far, we have gone to Alaska twice, Mexico thrice, Cuba, Nassau twice, the Virgin Islands twice and soon more journeys.

Since becoming a sailor a lot of folks have asked me the origins of drinking water on cruise ships. The perceptions people have presented to me have run the gamut of silly to ridiculous to outright impossible.

Still, people have real concerns about water safety when cruising. It is unfortunate that cruise lines do not say much about their water systems because they have a lot to boast about. Then again, they do not really detail their radar technology, engine specs, air, heating or electrical systems either. We just take it for granted, and rightly so, that everything is first-rate.

Almost to a man, everyone I have ever talked to about cruise ship water has some preconceived notion that the water is not safe, that it is likely full of contaminants and disease, and in general should be avoided.

Certainly, history gives reason for these beliefs to linger. In the old days, many ships would merely fill up huge onboard tanks with tap water, pressurize it with pumps and head out to sea, hoping they would not run out or encounter contamination. *Titanic* anyone? For sure the stories of cruising in the old days document illness due to contaminated ship water.

Then, as technology improved, some cruise lines looked for more efficient ways of providing water. At one point they carried onboard purification systems that would take toilet flushes and shower water and recycle it into drinking water. Sounds yucky, but it was safe.

The problem was that the reverse osmosis (RO) technology at the center of these systems often converted only 50% or so of the total wastewater back into drinking water. Accordingly, on those ships, there was the complication of collecting the waste of the water system and putting it into general waste tanks or releasing it into the open seas.

In addition, over the decades there were also ships that carried tanks of fresh water for drinking and then used seawater for toilets and showers. This created a problem of salt corrosion in the pipes, sinks, and drains, and the expense of complicated piping.

Hallelujah! Enter modern technology. Today the cost of desalinization RO has made it possible for cruise ships to take seawater and treat it, store it, and deliver it, as an endless supply to passengers. Unless there is equipment failure, your modern cruise ship will not likely run out of drinking water.

Not only is the amount available impressive, but so is the quality. Modern cruise ship water systems are virtually identical to those used to produce products such as Dasani, Nestle Water and Aquafina. The water is purified, minerals are added to make it taste good, and then it is sent to tanks of 1.5 million gallons or more. Once processed, sanitization technology such as ozonation, chlorination, UV light, and microfiltration are applied as a final treatment protocol. By the time you open the faucet in your cabin or take a drink in the restaurant or hallway fountain, you are being given a real treat of safe and delicious water.

On a final note, having inspected multiple cruise ship water systems, and having met with many cruise line water engineering staff, I can testify firmly that these professionals are among the finest, most well-trained, and intelligent water operators on earth.

Cruise ship water? Drink up, the water is fine!

Combatting drought by transporting water – A common belief is that we can solve a lot of the world's water needs simply by transporting water from places that have a lot of water to places that are starving for water. While this seems to make sense, a closer look demonstrates just how impractical this is other than for short-term emergencies.

- Chlorine dissipates with heat, motion, and time. Water transported any great distance is likely to arrive contaminated with bacteria and algae growing in the tank.
- The cost to transport a 9,500-gallon tanker of water 1,500 miles currently in the US is about $6,500. This translates to about 68 cents per gallon of water. As we mentioned in chapter 19, the average household uses about 350 gallons of water per day.

  This means that the average family would pay $238/day, roughly $7,140 per month, or $85,600/year for transported water. The true cost would be more because there would be additional markup applied by the companies that collected it from the tankers and then distributed it to the consumer. But even at 68 cents per gallon the economics are crushing. How could an average family afford this?

- Pipelines to deliver water from river or lake sources to communities in need, are something I advocate and promote, but are not immediately ready to build because constructing them involves the legalities of land rights, lease issues, and the fact that each community along the river or lake would be fighting to offer cheaper water than their neighbor.

- Then there is the question of who the river/lake water belongs to in the first place? Add to that the fact that depleting river and lake levels can have terrible environmental effects. These factors would need to be studied, agreed upon and legislated.

- Finally, and perhaps most importantly, we would have to find funding for what would be an immense investment in pipeline infrastructure, personnel, and oversight.

The ideas of permanently transporting drinking water by either tanker vehicle or pipeline have so many obstacles that neither offers a realistic long-term solution. Under the guidance of a national water plan a drinking water pipeline is certainly a long-term option that should be looked at.

I also want to make a clear distinction between the ocean water pipeline I advocate for putting a stop to our ravenous forest fires, and a drinking water pipeline. These are two vastly different projects, and the first idea is not only far more viable, but it is something that could and should be started today.

What is that pink stuff? – More than one of us has asked this question about pink deposits that we find in swimming pool filters, toilet bowls and sinks, and even home water filtration systems. Upon seeing this junk most people refer to it as a pink slime or mold.

This colorful bugger is a common but persistent airborne bacterium known as Serratia marcescens (SM).

Normally, in the air, SM is harmless to humans and animals. However, as with all organisms we do not want to purposely ingest it or get it into open wounds.

If your home or water system is contaminated with this stuff, a few precautions will help.

Understand, as we have mentioned many times now, bacterial contamination is a normal condition of water, so the prospect of eliminating SM growth completely may not be realistic. The good news is, there are things you can do to minimize your headaches with this pink monster.

- SM is resistant to halogens (chlorine, bromine, iodine, fluorine) and while a good scrubbing with household cleaners will remove the pink from your porcelain this will probably not eliminate the problem. To have greater success, use cleaners that contain peroxides and acetic acid (strong vinegar). These products will remove the traces on surfaces that you cannot see.

- Replace all clear plastic filter housings on your water system with opaque models. Clear plastic allows sunlight to strike the water inside the housing. Because many bacteria are photosynthetic, in other words stimulated to grow by sunlight, clear plastic housings will make your bacteria problems worse.

- If your area is damp and hot, find a way to cool it down and dehumidify it.

Air conditioning works wonders for minimizing this stuff.

On a final note, bacteria are normal. If your goal is to minimize bacteria then you must develop a solid, recurring protocol for sanitizing the system and creating an environment that will reduce contamination risks.

Water for CPAP machines – Like millions of Americans I am a CPAP machine user. The machine has brought peaceful sleep to me and probably saved my life. Strangely enough there is also a water-quality phenomenon that goes along with using a CPAP machine. Since these are two things of great interest to me, I want to share some important facts about this wonderful treatment technology.

A CPAP machine offers a water reservoir with an evaporation plate to allow the user to regulate the amount of humidity in the air the machine delivers.

Depending on where you live the water you put into a CPAP reservoir can cause problems. If hard water in the reservoir evaporates a nasty scale will be left behind on the evaporation plate. To alleviate this, medical professionals suggest the use of distilled water, which has little calcium in it and leaves no scale behind should the water all evaporate overnight. Likewise, distilled water is free of chlorine, which could irritate mouth, nasal and other tissues over the sleep process.

Distilled water, however, causes a hidden problem for the user. As we learned multiple times in the book, organisms will grow quickly in non-chlorinated water. For the user who buys distilled water in a gallon jug, there is a high probability that over time bacteria, viruses, algae and more will proliferate in the water. Personally, I would not keep an opened/recapped jug of distilled water for CPAP use more than a few days.

An interesting universally recognized issue with CPAP users is a higher-than-normal occurrence of tooth decay. Many journals point to dry mouth tissue as the culprit. I have a potentially different angle to consider, that angle being the distilled water.

If you remember, high-purity water does not exist in nature. As soon as it is created it begins to pull contaminants/chemicals from everything it contacts. Distilled water is a high-purity water and there is absolutely a possibility that the nightly 6–8-hour exposure of vaporized distilled water in the mouth of the CPAP user can pull calcium out of the user's teeth. There is probably enough science and problem here for the dental industry to look at this.

My own choice for CPAP water is Memphis tap water. We are blessed with super low hardness, low chlorine, and pretty much the lowest overall contaminant level in America. When I travel, I use Dasani or a similar brand. They are not hard, do not leave scale, are free of chlorine, and it is inexpensive enough that I can use a new bottle every night. However, as always, consult your doctor for advice on CPAP water.

The point here is that water, as it does on most things, has a great impact on the CPAP world.

Water for puppies and kitties – As a dog lover this might just be my favorite part of this book. If you love dogs and cats, this might be your favorite part, too.

It is important to change pet water often and clean his/her bowl. If you think your dog can just drink any water, think again. There are many organisms and parasites in water that can make your dog sick, cause blindness, organ failure and even death.[46] You might be shocked to find out that the water in your dog's bowl and even the bowl itself might be unhealthier and more dangerous than water out of a puddle, pond or lake.

When you first pour your pet's water from the tap, there is chlorine in it. Chlorine kills bacteria and other harmful organisms and keeps both pet water and people water safe. Unfortunately, chlorine begins to dissipate as soon as it is poured into a bowl. In a fairly short period of time there is not enough chlorine in the water to stop things like bacteria, viruses, algae, fungi from growing in the water and on the

46 https://www.aspca.org/news/pet-safety-alert-rising-dangers-blue-green-algae#:~:text=-Dogs%20can%20develop%20poisoning%20when,Seizures.

surfaces of the bowl. So here are some tidbits to keep your pet drinking safe, delicious water.[47]

- After 8–12 hours in the bowl, discard your pet's water and put in a fresh splash.
- After 24 hours your dog's bowl should be washed in hot water and detergent. Ideally Fido's water bowl should be put into the dishwasher daily.
- All bowls allow bacterial slime to form on the inside surface. If the inner bowl surface feels slippery to your touch, then the bowl is contaminated with bacterial slime. New chlorinated water will not kill this. If you suspect bacterial slime either run it through the dishwasher or clean it manually with a cleanser like Ajax. Make sure to rinse and wipe it thoroughly before putting new water in it.
- Porcelain bowls are very porous and can harbor and protect harmful organisms in all the microscopic nooks and crannies found in their surfaces. Porcelain should also be put in the dishwasher daily. Once a week it is a good idea to fill the bowl with hot white vinegar and let it sit for about 30 minutes. After you empty the vinegar, fill the bowl with hydrogen peroxide and let that sit for 30 minutes, afterward rinse and return to your pet. Porcelain is especially prone to contamination when left outdoors and I always avoid them. Stainless steel bowls are best.
- Plastic bowls are also good. You can scrub the heck out of them, and they love the dishwasher. Beware, however, that they scratch easily and the crevices that form can harbor bacterial and viral growth.[48]

---

47 https://www.youtube.com/watch?v=HsAiGq7wjzw&feature=em-upload_owner.

48 https://www.thisdogslife.co/do-you-know-about-the-dangers-lurking-in-your-dogs-water-and-food-bowls/.

- If you carry water for your pet in a container, remember, after 24 hours from being opened, that water also will be subject to contamination. Make sure you follow the guidelines for bowls with containers for bottles as well.

- If your community is under a boil alert, this must be applied to your pet's water as well. A pet is subject to many of the same contaminants, diseases and germs that make us sick.

Thoughts on America's water – You might have noticed that chapters on things like global water statistics, shortages, drought, and world contamination are absent from this book. Certainly, I could have added chapters on this, even an entire book. However, I chose not to address those issues because the purpose of my book is to help you gain a better, clearer grasp on the physical, chemical, mechanical and biological properties of water so that you can make better personal water decisions for you.

There are plenty of economic and political publications for those who want to enter those debates and activities. Still, there are a few global/national hot items I like to mention in my water journey. So, here is some food for thought.

- The American Society of Civil Engineers gave the US a "D" grade for the quality of its drinking water.[49]

- As many as 63 million people (1/5th of US), were exposed to potentially unsafe water more than once during the past decade (EPA).[50]

- Nearly 1/3 of US water systems reported service lines containing lead (American Water Works Association).[51]

---

49 https://www.waterworld.com/drinking-water/article/16223567/us-infrastructure-gets-near-failing-grade-in-new-asce-report.

50 https://www.usatoday.com/story/news/2017/08/14/63-million-americans-exposed-unsafe-drinking-water/564278001/.

51 https://www.awwa.org/Resources-Tools/Resource-Topics/Inorganic-Contaminants/Lead.

- In 2015, 18 million people were served by water systems with lead violations (NRDC).[52]
- Current EPA regulations and testing for lead are inadequate and leave consumers at risk for lead and other metal poisoning.[53]
- The American Water Works Association (AWWA) predicts that it will require $1 trillion to fix our failing infrastructure. It is highly unlikely that our government and taxpayers can afford that type of investment.[54]
- 70% of fresh water in the world is for agricultural purposes.

  19% of fresh water used in the world is used for industry.

  11% of the water used in the world is used by the public.[55]

Despite these numbers, according to the AWWA, there has been no significant reduction in water withdrawals from any of these groups in recent years. Now, I want y'all to think long and hard about this.

Several times throughout this book, while encouraging conservation and recycling, I have also boldly claimed that for a variety of reasons conservation activities will not have a major impact on our overall water shortages and needs.

I promise you that the overwhelming majority of "water experts" and conservationists will take great offense and exception to this and no

---

52 https://www.nrdc.org/experts/kristi-pullen-fedinick/whats-your-water-updated-analysis#:~:text=NRDC's%20new%20analysis%20of%20the,January%202015%20and%20March%202018.&text=EPA%20and%20health%20experts%20agree%20that%20no%20amount%20of%20lead%20is%20safe.

53 https://www.apmreports.org/story/2020/05/04/epa-lead-pipes-drinking-water.

54 https://www.infrastructurereportcard.org/cat-item/drinking_water/#:~:text=According%20to%20the%20American%20Water,over%20the%20next%2025%20years.&text=Over%20%24105%20billion%20is%20needed%20for%20water%20funding.

55 https://www.mckinsey.com/business-functions/sustainability/our insights/water-a-human-and-business-priority#.

doubt blast me. I can hear them now: "You charlatan, how dare you marginalize conservation?" My retort is simply that statistics do not lie and in fact support me.

For all our altruism, idealism, best intentions, and conservation efforts each and every day we suffer every new and increasing contamination and water shortages. So, what is the problem here? What are we missing? What we are missing is that it is not climate change, drought or pollution that are causing our water shortages. It is instead our lack of vision, planning and understanding of water. It is our own poor behavior.

Let me be perfectly clear. With 71% of the earth being covered in water, albeit 97% seawater, the belief that you and I need to conserve, ration and hoard water is not a statement of responsibility, it is an admission of buffoonery.

It is not the earth's fault that man overdevelops areas with known water limitations. It is not climate change that causes states like California, which have an abundance of rainfall and snowmelt, to run out of water due to not building more storage capacity. It is not drought that prevented us from building ocean pipelines that could have been quenching our wildfires for decades.

Most of these water problems are self-inflicted. The ones that are not, we could have eliminated if we had planned and made water a priority.

The great water grabbers – Some of the fiercest battles ever fought, both physically and legally, have taken place over water rights. I am here to assure you that as water becomes scarcer and more expensive these types of battles will become much more prominent.

At the root of all these battles, always, is a person or an entity that I call a water grabber. In of and by itself the term *water grabber* is neither a positive label nor a negative label. Instead, it is merely the name I use for the person or entity that is responsible for the initiation of a water battle, or if you will, a grab for water.

In an ideal world there is enough water for all of us, and when there is not, we would like to think that those who have it will make sure others do as well. But such is seldom the case and I have lived to see brothers, close friends, and communities go to war over water.

Imagine that a community along the Mississippi River decided to build a hydroelectric plant. Doing so would dramatically reduce the flow to the next communities downstream, changing the ecosystem, recreation and even transportation dynamics. But doesn't the community have the right to do what they want with their water?

I read a story, years back, of a landowner who was selling water out of an aquifer that was under his farm to one of the large beverage companies to make bottled water. At first glance this did not strike me as being a big deal, and at first, I thought it not even newsworthy. However, as I read on, I learned that this man's aquifer also supplied additional pockets of water from which nearby communities obtained their drinking water. A legal battle ensued, and I confess to not following through to see how it was resolved, but this is yet another example of "Whose water is it?"

The state of Mississippi filed a lawsuit against the state of Tennessee for an enormous multimillion-dollar claim maintaining that the action of Tennessee pulling water out of its own aquifer was causing water to be pulled from the nearby and possibly adjoining Mississippi aquifer. I do not believe this issue has been ruled on as of this writing, but it is for sure a mess.

More than one citizen has gone to war with the federal government that dared to make environmental claims on water bodies contained within the landowner's property lines.

So, whose water is it? If the water flows to us first, do we have the right to deprive others who live downstream? Should industry, farms and even simple homeowners be restricted on how much water they can pull out of their pipes, faucets, and pumps?

If we were to build a water pipeline would the community where the pipe went into the lake, ocean or river have the right to be compensated for the water taken? What about others who rely on the same body of water?

These are not easy questions to answer, but I assure you in the coming times we will see the emergence of a whole new culture of water grabbers, whether they be individuals, government agencies and businesses who, for one reason or another, noble or opportunistic, will create great havoc in the determination of water rights. Do not be surprised to see this come to your own neighborhood!

More on water jobs

When I accepted my first job as water sales engineer, I had no idea just how vast the water industry was. I think most people still do not realize it. For those looking for jobs or a rewarding career you might not want to look any further than the water industry.

- Forbes estimates that fixing America's failing water infrastructure will create 1.9 million jobs.[56]
- Jobs in the water sector offer above-average pay and strong advancement opportunities.[57]
- There is an untold bounty of career opportunities for sales engineers that solve water problems in industry. Companies

56 https://www.forbes.com/sites/petergleick/2011/10/05/jobs-and-water-for-america/?sh=-7426b9c26c7f.

57 https://study.com/articles/jobs_that_involve_the_water.html.

> like Koch, Evoqua, Culligan, Calgon, Coke, ITT, GE, and many more have a constant and growing need for individuals with science degrees who want to fix water problems. These are all well-paying jobs with stability and longevity. An Internet search for water sales engineer jobs may very well knock your socks off.

In the world of manufacturing water is a crucial lifeblood to success—water for boilers and cooling, pure water to manufacture electronics, high-purity water to rinse metals in the automotive/marine/appliance, and of course water for food and beverage. Then there is the need for wastewater professionals as well as those who work on water recycling. Hospitals have soft water systems and pure water systems for labs, dialysis, central sterile services and more.

There is nary a manufacturing plant that does not rely on some form of water treatment and those who operate those systems.

As a closing thought, we all know that the world's water problems are increasing and will continue to do so. All these things make the water industry a perfect vocation with which to provide for yourselves and your loved ones.

Viva Las Vegas

My favorite place on earth, Las Vegas is a perfect example of how powerful water is, and how it feeds our lives. Most people associate the discovery and boom of Las Vegas with gambling and credit Bugsy Siegel for birthing this American icon. The truth, however, is that it was water and not gambling that was responsible for the birth of Las Vegas, and the progressive treatment water is the reason Las Vegas continues to thrive today.

The name Las Vegas was bestowed on this area by a man named Rafael Rivera, an explorer who was a member of a party led by the Spanish explorer Antonio Armijo. The area was previously listed on maps simply as Vegas, but somewhere around 1830 Rafael renamed the area Las Vegas, which translates to "The Meadows."[58]

The reputation of this area for having plentiful artesian wells spread to the many traders who traveled along the Old Spanish Trail from New Mexico to Los Angeles. Traders would take a diversion to refresh with water. Soon the reputation spread to others who brought food and drink, livery, lodging and other services to the area. The underground flow of these wells also ran to the Colorado River.

Gambling, no doubt, arose from the visiting traders needing something to do. Bugsy Siegel and the glitz of the strip came to the area many years later. If you are like me and love Vegas, the next time you visit, share this story with someone new.

My favorite water organizations

If your interest in water is piqued, and I hope it is, the following entities are the absolute best places I can think of to help you learn even more about water. Check them out and enjoy yourselves.

- The Water Quality Association (WQA), which can be found at wqa.org. The WQA is an absolute crowd-pleaser and leader of consumer, commercial and industrial water culture, and technology. They offer certification classes for those wanting professional status as well as product reviews and an excellent annual trade show where water professionals gather from all over the globe to share ideas and products. Perhaps the most wonderful feature of the WQA is that no matter what your

58 https://www.britannica.com/place/Las-Vegas-Nevada/History#ref983345.

level of water understanding, their programs and publications have something to offer.

- The American Water Works Association (AWWA), which can be found at awwa.org. The AWWA boasts the largest membership of water professionals, and members. The group publishes a voluminous amount of material that applies to water technology and trends. Through a series of events, including an annual national conference, the AWWA offers a wealth of information to those seeking water knowledge. Their "State of the Industry" report is a comprehensive report on infrastructure, personnel, investment, contamination, employment, and virtually anything else a person could want to know about water. This is a place of higher learning and the newbie might feel a bit overwhelmed at first. However, for the individual seeking the highest understanding of water, this group is a humdinger.

- The Association of Water Technologies (AWT), which can be found at AWT.org is another organization that promotes water professionals. However, through seminars, teaching materials and regular events this is a group that can help educate the average consumer on the wide world of water technology.

- The United States Geological Service, USGS, which can be found at USGS.org. By their own description, the United States Geological Survey is a scientific agency of the United States government. The scientists of the USGS study the landscape of the United States, its natural resources, and the natural hazards that threaten it. The organization's work spans the disciplines of biology, geography, geology, and hydrology. If your interest is tracking drought, water availability and stresses to our water resources, the USGS is a must-see.

- Water.Org, which can be found at water.org, is the world's most famous and most successful group in bringing safe water to the world. Through scientific contributions, volunteerism and benevolent financial support, this organization continues to grow and educate the world. The organization is not involved so much in local water issues but rather global ones. Still, this is a great place to learn about coming water trends.

If you research water organizations online you will be rewarded with a mind-numbing return of other water entities, some of them new, some of them longstanding. Rest assured, there is more water information available than a person can read and digest in a lifetime. By not including other organizations, I by no means meant to indicate that other groups are not valuable and worthy of our time. However, given the intent of this book, I believe the five groups I listed will provide ample avenues for expanding the readers' water education.

22.
Final Thoughts

I have wanted to write this book for 10 years, and now that it is finished, I can tell you that it was a difficult and frustrating process. Despite my passion for water, and an equally sincere passion for helping others understand it, the science of water is, as you have now learned, confounding, elusive and complex. I am sure that you noticed on more than a few occasions when introducing a new principle or idea, the explanation involved bringing up things we learned in previous chapters.

As an example, our initial conversation about non-chlorinated water came back to us throughout the book. Whether it was bottled water, home filters, RO systems and more, to explain the new concept we had to revisit chlorine. And so, this is the world of water. It is a science of building blocks, repetition, discipline, and as I mentioned early in the book, it is a science in which answers are seldom straightforward.

At any level of expertise, it is a difficult topic to get one's hands around. My lifelong labor of love remains a pain in my ass.

As simple, universal, beautiful, refreshing, healing and delicious as water is, it has so many hidden faces, natures, behaviors, and scientific properties that it eternally remains the ultimate jigsaw puzzle. Just when we think we know it all, water humbles us with a new twist that we never saw before.

Here is the challenge we all face in our quest for safe water: when we are asked, "What is the best home water filter," we cannot reply intelligently until we gain specific information, such as what contaminants are to be removed? how much water per day do you need? how fast do you need it? what is the pH, hardness, alkalinity? and on it goes.

Unfortunately, often, when we respond in this detailed manner, the questioner quickly loses interest and after a few words moves on down the line to find the expert who will give them that simple, straightforward answer to their questions.

It is this universal impatience and demand for simple water answers that continues to render us vulnerable to bogus water products and irresponsible water behavior by our leaders. It is this demand for simple answers that stands between you, me, and safe water.

Soberingly, it is this exact rush to simple answers that caused the deadly contamination of water in Flint, Michigan. Standard water practices dictated that when the Flint water source was switched, there was a need for a new analysis, new treatment prescription and new protocols. Instead, they were impatient, they did not want to be confused with complicated water details and they simply strapped the old treatment recipe onto the new water and in the end jeopardized the health and safety of the entire community.

So now on to my very final thoughts.

Some think me mad, but I routinely teach that water is one of the most dangerous and deadly ingredients of our existence. It surely was in Flint, and a simple Internet search will direct you to countless other drinking water horrors that could have been avoided if someone just would have taken the time to do the homework.

When it comes to your own water journey, I want you to do the homework and ask the tough questions. I want you to push others to the point of frustration in answering your questions and providing the water solutions you seek.

My book is by no means a complete, everything-you-need-to-know-about-water document. But it is the best tool available to get you on your way to gaining your own command of water.

As best I knew how, I compartmentalized all the ingredients of my career and pared them down into logically arranged and hopefully easy-to-understand chapters. Hopefully, my book will not simply be a read once and put down, but instead it will become your water companion and good friend for years to come. If that happens then I will consider this book to be my greatest success.

I wish you all, safe water.

Glossary of Terms

TDS	Total Dissolved Solids
RO	Reverse Osmosis
EPA	Environmental Protection Agency
PFAS	Per Fluoro Alkyl Substances
WFI	Water for Injection
THM	Trihalomethanes
SVOC	Semi-Volatile Organic Compounds
LSI	Langelier Saturation Index
GAC	Granular Activated Carbon
CA	Cellulose Acetate
TFC	Thin Film Composite
SDI	Silt Density Index
BWF	Back-Washing-Filter
NSF	National Sanitation Foundation
WAC	Weak Acid Cation (Ion Exchange Resin)
IXR	Ion Exchange Resin
SAC	Strong Acid Cation (Ion Exchange Resin)
MBR	Mixed Bed Resin
CR	Cation Resin
AR	Anion Resin
GPM	Gallons Per Minute
GPD	Gallons Per Day
PPM	Parts Per Million
CFL	Compact Fluorescent Light